OXFORD MEDICAL PUBLICATIONS

BASIC CLINICAL PHYSIOLOGY

BASIC
CLINICAL PHYSIOLOGY

J. H. GREEN
M.A., M.B., B.Chir. (Cantab.)
Ph.D. (Lond.), A.R.I.C.

Professor of Physiology, University of London
at the Middlesex Hospital Medical School

SECOND EDITION

LONDON
OXFORD UNIVERSITY PRESS
NEW YORK   TORONTO

Oxford University Press, Ely House, London W.1

GLASGOW NEW YORK TORONTO MELBOURNE WELLINGTON
CAPE TOWN IBADAN NAIROBI DAR ES SALAAM LUSAKA ADDIS ABABA
DELHI BOMBAY CALCUTTA MADRAS KARACHI LAHORE DACCA
KUALA LUMPUR SINGAPORE HONG KONG TOKYO

ISBN 0 19 263326 0

© Oxford University Press 1969, 1973

First published 1969
Second Edition 1973
Reprinted (with revisions) 1975

Reproduced and printed by photolithography and bound in
Great Britain at The Pitman Press, Bath

CONTENTS

PREFACE

This book has been written as an introductory text for students of pure and applied physiology who wish to relate the basic concepts of modern human physiological knowledge to clinical medicine in its widest sense, but who do not wish to be overwhelmed, at this stage, by complex theoretical considerations or advanced chemistry, physics or mathematics. It is hoped that this book will also prove useful as a short text for the rapid revision of the principal concepts of the subject.

This book has been written for students of the Medical and Nursing Faculties, and for students of the professions supplementary to medicine who have, or are about to have, contact with patients and who may be having introductory lectures in pathology and medicine at the same time as their lectures in physiology.

With the current trend towards integration between subjects, it is hoped that this book will prove useful to those interested in bridging the gap between the pre-clinical and clinical subjects.

This book should be of particular interest to nurses taking the 1969 syllabus of subjects for the Certificate of General Nursing, and especially those who are seeking a higher qualification such as the Diploma in Nursing or the Nursing Tutor's Diploma.

It is hoped that it will enable students of all grades to relate their physiological knowledge to the clinical cases seen in the wards, and will encourage the consultation of more advanced texts for the further study of topics of special interest. References have not been included as they will be found in these books.

It is realized that many clinical conditions have, at the present time, an unknown physiological background, but such conditions are becoming fewer as further research reveals more information concerning the way the body functions.

I should like to thank all those who have made helpful comments and criticisms which have been of great assistance in the preparation of the new edition of this book.

In this edition six new figures have been added and a further eight figures have been amended. In response to requests, the sections on fluid and electrolyte balance, the cardiovascular system, the liver and digestion have been expanded. New or additional material has been included on: cerebrospinal fluid, tissue typing, heart rate, functions of atria, angina pectoris, blood pressure, determination of cardiac output, fainting, shock, exercise, blood, lung volumes, carbon monoxide poisoning, respiratory obstruction, textured foods, nitrogen balance, lactose intolerance, disorders of digestive function, monoamine oxidase inhibitors, vitamin D, prostaglandins, diuretics, molarity, hypothalamic releasing factors, thyroid, lumbar puncture, dioptres, audiometry and electromyography. In addition numerous alterations and amendments have been made to the text and in the arrangement of the material presented.

The Middlesex Hospital Medical School   J.H.G.
London, W.1

1. INTRODUCTION

Physiology is the study of how the body functions. In its widest sense physiology is concerned with the whole of the animal and vegetable kingdoms, but in this book we are concerned with man.

Physiology is an experimental subject, and information is obtained, wherever possible, by making objective measurements. Thus, although placing the hand on a person's brow may suggest that the person is feverish, a clinical thermometer will settle the matter by measuring the patient's temperature (or will it? See Chapter 14).

This is a book about a hypothetical man who is the average of all, and how his body functions. The numerical values given to the physiological constants of this standard man will act as a guide to the value expected in any normal individual.

The body is built of cells which are grouped together in organs. These organs form systems of body function. Thus the heart, blood vessels and the blood form the **cardiovascular system.** The lungs, and air passages, together with the respiratory muscles, form the **respiratory system.** The **digestive system** converts the food taken in by mouth into a suitable form for growth and repair of tissues and for the production of heat and energy. Waste products are excreted from the body via the **urinary system.** This system consists of the kidneys which form urine. This is passed along the ureters to the bladder. After temporary storage the waste products pass out dissolved in water via the urethra as urine.

The body functions are controlled by the **control systems** and these consist of two parts: the endocrine glands produce hormones or 'chemical messengers' which circulate via the blood stream to act on a distant organ whereas the nervous system conveys information along nerves in the form of coded nerve impulses. The peripheral nervous system is controlled by the central nervous system which consists of the brain and the spinal cord. The body is kept informed of the external environment by the special senses of sight, smell, hearing and taste.

This subdivision of the body into systems is for convenience only and it must be remembered that all the systems are interconnected.

Cardiovascular System

The first system to be discussed is the **cardiovascular system.** This is the transport system of the body [FIG. 1]. It carries oxygen from the lungs to the tissues. It carries carbon dioxide from the tissues to the lungs so that it may be excreted in the expired air. It carries food from the digestive tract to the cells of the body so that these cells may extract nutrients for growth and energy. Thus if the blood supply to a part of the body is impaired, the

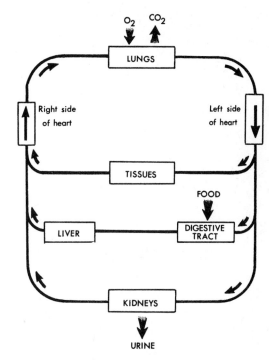

FIG. 1. The cardiovascular system forms the transport system of the body. It carries:

1. Oxygen and carbon dioxide.
2. Food.
3. Waste products to kidney for excretion in urine.
4. Hormones.
5. Heat.

repair of damaged tissues will be delayed. It carries waste products from the cells round to the kidneys for excretion in the urine. It carries hormones from the endocrine glands to the other organs of the body. It also carries heat from the parts of the body where heat is produced to the skin so that the surplus heat can be given off.

2. THE HEART AS A PUMP

The heart is the pump which circulates the blood round the body. It is in fact two pumps, the left side of the heart pumps blood from the lungs to the tissues, and the right side of the heart pumps the blood which has returned from the tissues through to the lungs [FIG. 2].

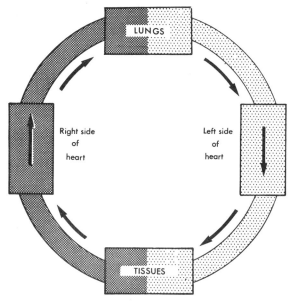

FIG. 2. The circulation of the blood. The pale shading indicates the oxygenated blood being pumped by the left side of the heart from the lungs to the tissues. The dark shading shows the partly deoxygenated blood which is returning via the veins and is being pumped by the right side of the heart to the lungs.

The blood which arrives at the left side of the heart is bright red in colour having picked up oxygen during its passage through the lungs [FIG. 2]. It is then pumped by the left side of the heart into the aorta, from which it passes via the arteries and arterioles to the capillaries. As the blood passes through the tissue capillaries, it gives up its oxygen and becomes dark-blue in colour. This dark blood returns to the right side of the heart which pumps it via the pulmonary artery to the lungs. In the lungs the blood picks up the oxygen, and its colour changes once again to bright red. It returns to the left side of the heart via the pulmonary veins.

Each side of the heart consists of two chambers, the atrium and the ventricle [FIG. 3]. It is the ventricles that play the major part in circulating the blood.

Each time the heart beats, each ventricle pumps out about 70 ml. blood. This volume is termed the **stroke volume**. The heart beats about 70 times a minute. This is termed the **heart rate**.

Multiplying these two together will give the volume of blood pumped out by each ventricle per minute. This is termed the **cardiac output**.

At rest, each ventricle will pump out 70 × 70 ml. blood per minute. This will equal 4,900 ml./min., which is near enough 5 litres per minute.

The fact that the cardiac output is equal to the heart rate multiplied by the stroke volume may be represented by the equation:

$$C.O. = H.R. \times S.V.$$

where C.O. stands for cardiac output, H.R. stands for heart rate, and S.V. stands for stroke volume.

We have about 5 litres of blood in the body, and the fact that the output of each ventricle is 5 litres per minute means that, on the average, the blood must be

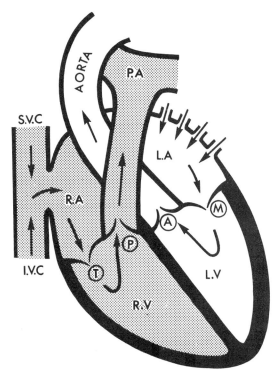

FIG. 3. The heart valves. The blood returning to the heart via the superior vena cava (S.V.C.) and the inferior vena cava (I.V.C.) passes to the right atrium (R.A.) and passes through the tricuspid valve (T) to the right ventricle (R.V.). The blood is pumped via the right ventricle through the pulmonary valve (P) into the pulmonary artery (P.A.). Blood which returns via the four pulmonary veins (arrows) to the left atrium (L.A.) passes through the mitral valve (M) to the left ventricle (L.V.). The left ventricle pumps the blood out through the aortic valve (A) into the aorta.

circulating round once every minute. Some blood in fact is circulating more rapidly than this and some blood is circulating more slowly. But on the average it will take one minute for a red blood cell to leave the left ventricle to pass via the arteries to the tissues, to return via the veins to the right ventricle, to pass via the pulmonary artery to the lungs, and then to return via the pulmonary veins and left atrium to the left ventricle again.

In exercise, both the heart rate and the stroke volume increase, and as a result the cardiac output is greater than at rest. For example, in very severe exercise, the heart rate may increase to 200 beats/minute, and the stroke volume may increase to 150 ml./beat so that the cardiac output now becomes 200 × 150 = 30 litres/minute.

It will be noted that with only 5 litres of blood in the body, a cardiac output of 30 litres per minute of blood means that the same blood is going round six times every minute.

CARDIAC CYCLE

The contraction phase of a chamber of the heart is termed systole (pronounced *sis'to.lee*). The relaxation phase is termed diastole (pronounced *dye.ass'to.lee*). At rest, ventricular systole lasts three-tenths of a second (0·3 second). Ventricular diastole is longer and lasts five-tenths of a second (0·5 second) [FIG. 4]. Diastole

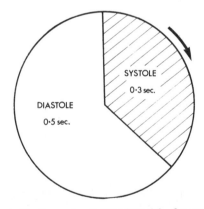

FIG. 4. Cardiac cycle. Systole lasts 0·3 of a second, diastole lasts 0·5 of a second making a total of 0·8 of a second for the complete cycle. Each systole is followed by a diastole which is followed by another systole and so on.

is then followed by the next systole which is followed by the next diastole and so on. The sequence of one systole followed by one diastole is termed a **cardiac cycle**. It lasts eight-tenths of a second.

Unlike skeletal muscle, cardiac muscle has no long periods of rest and has to continue its activity throughout the whole of life. The only rest period it has is during diastole. When the heart rate speeds up, as it does in exercise and emotional excitement, the whole

cardiac cycle is completed in a shorter time. This reduction in the duration of the cardiac cycle is mainly at the expense of diastole i.e. the rest period. A persistently fast heart rate reduces the over-all time available for diastole.

There is a wide variation in the **heart rate** of young healthy adults at rest. This rate may be as low as 40 beats a minute or it may be as high as 90 beats a minute. With exercise and emotional excitement the rate increases to a maximum of 210 beats per minute. By middle-age this limit has fallen to about 160 beats per minute. The factors regulating heart rate are discussed on page 28.

Atrial systole lasts only one-tenth of a second and the remaining seven-tenths of a second correspond to atrial diastole.

HEART VALVES

Each ventricle has a valve at its inlet, and a valve at its outlet [FIG. 5].

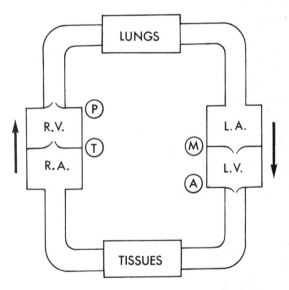

FIG. 5. Heart valves. The blood is circulated by the contraction of the ventricles. Each ventricle has a valve at its inlet and a valve at its outlet. [See also FIG. 3.]

(M) = mitral valve, (A) = aortic valve, (T) = tricuspid valve, (P) = pulmonary valve, L.A. = left atrium, L.V. = left ventricle, R.A. = right atrium, R.V. = right ventricle.

The inlet valves are termed the **atrioventricular valves**. On the left side it is also known as the **mitral valve**. On the right side it is known as the **tricuspid valve**. The mitral valve has two cusps; the tricuspid valve has three cusps.

The outlet valves have three cusps and are known as the **semilunar valves**. The valve on the left side of the heart is also known as the **aortic valve**, the valve on the right side of the heart is known as the **pulmonary valve**.

The heart valves are structures which allow the blood to flow in one direction only. When there is any tendency for the blood to flow in the other direction, the heart valves close. Valves are passive structures, that is, they do not contain any muscle tissue. As a consequence diseased valves may be replaced· by valve transplants (homografts), by valves from animals (heterografts) or by mechanical valves (prostheses).

With the onset of ventricular systole, the atrioventricular valves shut and shortly afterwards the semilunar valves open. At the end of ventricular systole the semilunar valves shut and shortly afterwards the atrioventricular valves open.

HEART SOUNDS

The closure of the mitral and tricuspid valves is associated with a sound that can be heard by placing one's ear to the chest, or more conveniently by using a stethoscope. It is termed the **first heart sound**. Through the stethoscope it is like the word 'lub' spoken very softly. This sound marks the onset of ventricular systole.

The closure of the semilunar valves (aortic and pulmonary) causes the **second heart sound**, which is like the word 'dup' spoken softly. This marks the end of ventricular systole and the start of ventricular diastole [FIG. 6].

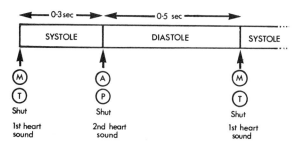

FIG. 6. Heart sounds. With the onset of systole the mitral (M) and tricuspid (T) valves shut. This gives the first heart sound. At the end of systole the aortic (A) and pulmonary (P) valves shut. This gives the second heart sound. Systole occurs between the first heart sound and the second heart sound; diastole occurs between the second heart sound and the next first heart sound.

Since the interval between the closure of the atrioventricular valves (first heart sound) and the closure of the semilunar valves (second heart sound) is shorter than the interval between the closure of the semilunar valves and the next closure of the atrioventricular valves, the beating of the heart has a characteristic rhythm *lub-dup*—pause—*lub-dup*—pause (1–2—pause—1–2—pause).

The heart sounds enable events in the cardiac cycle to be timed. Any event which occurs in the interval between the first and second heart sounds is occurring in *systole*. Any event which occurs between the second heart sound and the next first heart sound is occurring in *diastole* [FIG. 6].

When the heart rate speeds up and diastole becomes shorter, the characteristic *lub-dup*-pause-*lub-dup*-pause rhythm may be lost, and it may be replaced by a rhythm in which the heart sounds appear at regular intervals, giving a *tup-tup-tup-tup* rhythm. In such a case it may be difficult, when listening to these heart sounds, to decide which is the first heart sound, and which is the second heart sound. In such a situation, the carotid pulse is simultaneously palpated, and the sound which corresponds in time to the carotid pulse will be the first heart sound.

With very fast heart rates, such as are found with severe exercise, diastole may become shorter than systole so that the rhythm reverses and becomes a very fast *dup-lub*—pause—*dup-lub*—pause (2–1—pause—2–1—pause).

Phonocardiography

Using a microphone attached to the chest, a graphical recording of these heart sounds can be made. It is termed a phonocardiogram [FIG. 7]. On such a record it is often possible to detect two additional heart sounds. The third heart sound, which occurs shortly after the second, corresponds to the rushing of blood into the ventricles during early diastole. The fourth heart sound, which corresponds to atrial systole, occurs shortly before the first heart sound.

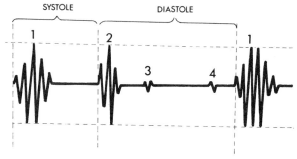

FIG. 7. Phonocardiogram. The heart sounds are recorded using a microphone applied to the chest. The first heart sound is produced by the closure of the mitral and tricuspid valves. The second heart sound is produced by the closure of the aortic and pulmonary valves. In addition a third heart sound and a fourth heart sound may be recorded.

Silent Blood Flow

When blood is flowing through a blood vessel, the flow is laminar [FIG. 8, *upper*]. The outer layers of blood close to the vessel wall are moving very slowly, the stream of blood nearer the centre of the vessel is moving more rapidly, and the fastest speed is found in the centre of the blood vessel. The blood in the centre is sliding with reference to the blood slightly to one side and this is sliding with reference to the blood next to it and so on to the edge. The blood flow is in layers or laminae and this has given the name to the **laminar flow.** (Multiplywood is called laminated-wood because it is made of a series of layers.)

An important point about this type of blood flow is that it is silent. No sound is produced, for example, as the blood flows through the radial artery. This fact can be confirmed by placing the wrist to the ear. No sounds of blood flow can be heard. Similarly, under normal circumstances, no sounds are heard if a stethoscope is placed over an artery (see recording of blood pressure, p. 24).

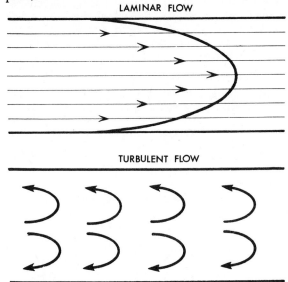

LAMINAR FLOW

TURBULENT FLOW

FIG. 8. *Upper.* The blood flow through a blood vessel is laminar. This type of flow is silent.

Lower. When the blood flows at high velocity through a narrowing in the circulation the flow becomes turbulent. This produces vibrations which may be audible or may be palpable.

Noisy Blood Flow

However, if when the speed of blood flow exceeds a certain value (known as the Reynolds number) eddy currents appear, the flow ceases to be laminar and becomes turbulent [FIG. 8, *lower*]. The vessel wall will now vibrate and this will give rise to sounds. This is particularly likely to occur if there is a partial obstruction to a blood vessel, since the blood will flow past the obstruction with a high velocity. This fact is made use of in the recording of blood pressure by the Korotkov sounds method [p. 24].

HEART MURMURS

The flow of blood through the heart is normally silent, but, should a valve fail to open completely, it will act as a partial obstruction. The blood will speed up and reach a high velocity as it passes through the narrow opening. This will lead to turbulence and to sounds and

vibrations in the chest. These sounds are termed **heart murmurs** and they can be heard by applying a stethoscope to the chest wall.

If the vibrations have a low frequency they may not be audible but can be felt if the hand is applied to the chest wall. Such palpable vibrations are termed **heart thrills.**

The condition of valve narrowing so that it does not open completely is termed **stenosis.**

An alternative cause of heart murmurs is the failure of a valve to close completely thus leaving a small aperture through which the blood flows very rapidly in the reverse direction to normal. This state of affairs is termed **incompetence** or **regurgitation.**

Murmurs Due to the Aortic and Pulmonary Valves

The murmurs, which are produced when the blood is flowing through the narrow part of the circulation are loudest when the blood flow reaches its greatest velocity. The phase of the cardiac cycle at which the heart murmur is heard is thus determined by the phase at which the blood is flowing rapidly through the obstruction.

With aortic stenosis, the high velocity occurs during ventricular systole and hence the murmur is produced during systole. It is termed a **systolic murmur.**

With aortic regurgitation the blood flows back at a high velocity from the aorta into the ventricle during ventricular diastole and this gives rise to a **diastolic murmur.**

On the phonocardiogram systolic murmurs are seen between the first and second heart sounds, whereas diastolic murmurs are seen between the second and the next first heart sound.

Murmurs caused by the pulmonary valve have the same timing as those caused by the corresponding lesion in the aortic valve.

Murmurs Due to the Mitral and Tricuspid Valves

Stenosis of the mitral or tricuspid valve will cause a murmur during diastole, and if the atria are still contracting, and not fibrillating, this murmur will reach its peak during the atrial contraction phase at the end of the ventricular diastole. The murmur of mitral stenosis is thus a crescendo **diastolic murmur** reaching its peak at the end of diastole.

The murmur of mitral incompetence (or mitral regurgitation) will be a murmur which arises during ventricular systole when some of the blood, instead of being pumped out into the aorta, will leak backwards through the incompetent mitral valve into the left atrium. This is thus a murmur which is occurring in systole, and is termed a **systolic murmur.**

VALVE AREAS

The lungs act as a poor conductor of sound, and the heart sounds and murmurs can be heard best when there is no lung tissue between the heart and the chest wall. The sounds produced by the closure of the pulmonary and aortic valves can be heard most clearly in the pulmonary and aortic areas in the region of the second ribs. The pulmonary artery comes close to the surface in the second left interspace (2 L.I.S.) [FIG. 9] whereas the aorta comes closest to the surface under the second costal cartilage on the right side of the sternum.

It will be noted that rib spaces are named with reference to the rib immediately above. Thus the second interspace lies below the second rib and between the

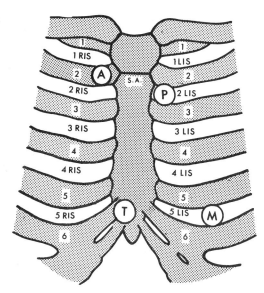

FIG. 9. Valve areas. Positions on the chest wall where the closure of the four valves is clearly heard. A = aortic valve. P = pulmonary valve. T = tricuspid valve. M = mitral valve.

If the sternum is palpated, by running the finger down the midline, the sternal angle (S.A.) will be felt as a ridge about 2 in. (5 cm.) below the jugular notch. This sternal angle (angle of Louis) is the junction between the manubrium and the sternum. It is an important landmark since the second rib (costal cartilage) joins the sternum at this level. It enables the second rib to be accurately located.

Interspaces are named according to the rib above. The second interspace lies between the second and the third ribs.

The mitral valve is heard most clearly at the apex beat which usually lies in the fifth left interspace (5 L.I.S.) 3–3½ in. (7·6–8·9 cm.) from the midline. This corresponds to a line through the middle of the clavicle (midclavicular line).

second and third ribs. The second rib joins the sternum at the sternal angle (manubrio-sternal junction) which is felt as a ridge about two inches below the jugular notch.

It should be remembered that the pulmonary and aortic valves produce the second heart sounds and when listening over the pulmonary and aortic valve areas only the second heart sounds may be heard.

The closure of the tricuspid valve and the murmurs produced by this valve can usually be heard most clearly at the lower borders of the sternum at the sixth costal cartilage on the right side.

APEX BEAT

As the heart contracts it swings forward and hits the chest wall in the fifth interspace about 3–3½ in. (7·6–8·9 cm.) from the midline, and in many subjects this impact can be seen as the **apex beat**. It may be felt by applying the flat of the hand to the chest.

The position of the apex beat from the midline corresponds approximately to the mid-clavicular line, that is, a line drawn vertically downwards through the middle of the clavicle. The apex beat is in most subjects the best place to listen to the mitral valve closure and to sounds produced by this valve.

The third and fourth left interspaces (3 L.I.S. and 4 L.I.S.) at the sternal border are places where sounds from all four valves can often be heard clearly.

FUNCTIONS OF THE ATRIA

The blood which returns in a steady fashion along the veins [FIG. 10] enters the right side of the heart, and during diastole when the atrioventricular valve is open, this blood passes directly into the right ventricle and

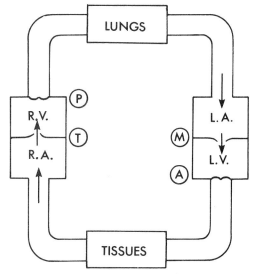

FIG. 10. Functions of the atria. Most of the blood returning to the heart during ventricular diastole passes straight through the atria and atrioventricular valves and collects in the ventricles. Shortly before the next ventricular systole the atria contract and transfer additional blood to the ventricles, thus completing the ventricular filling. This atrial systole improves the efficiency of the heart as a pump, but is not essential for life.

Blood returning to the heart during ventricular systole finds the atrioventricular valves closed. This blood collects in the atria.

collects there. During ventricular systole, the atrio-ventricular valve is shut, and the blood returning to the heart can no longer enter the ventricle. Instead it accumulates in the right atrium. The right atrium is thus a 'waiting-chamber'. It acts as a reservoir to hold the blood which is returning to the heart during ventricular systole and which can no longer enter the ventricle.

It is possible for circulation to be maintained without any atrial systole. The atrial systole, which occurs in the normal beating heart, improves the filling of the ventricles with blood. By the time the end of diastole has been reached, the ventricles are about three-quarters full of venous blood which has returned to the heart. Then in the last tenth of a second of ventricular diastole, the atria contract and transfer their contents to the ventricles. This puts in the last one quarter of the blood.

Exactly the same thing is happening on the left side of the heart, as on the right, but in the case of the left side of the heart, the blood is returning from the lungs.

The atria are thus acting as a booster pump, which provides the ventricles with more blood to pump out at the next heart beat.

3. CARDIAC MUSCLE AND THE ELECTROCARDIOGRAM

Cardiac Muscle

The heart beats because it is made of cardiac muscle. Each cardiac muscle cell has the property of **rhythmicity**, that is, it alternately contracts and relaxes or **beats**. The cardiac muscle cells also have the property of **conduction**. Each cardiac muscle fibre branches and comes in very close contact with several other cardiac muscle cells. When one cardiac muscle cell starts to contract the contraction wave spreads to the neighbouring cells and they also contract. In this way, a contraction-wave starting at one part of the heart spreads rapidly to all the other cardiac muscle cells.

Origin and Propagation of the Cardiac Impulse

In the amphibian, such as the frog, the heart beat can be seen to originate in the **sinus venosus**. This is an additional chamber found between the veins and the right atrium.

In man the sinus venosus is no longer present. All that remains is the **sinu-atrial node** (S-A node), which consists of a group of cells found in the right atrium close to the entry of the superior vena cava.

The heart beat in man originates in the S-A node. This node acts as the **pacemaker** of the heart, and the rate of beating of the S-A node determines the heart rate. Because the S-A node beats at a faster rate than the other cardiac muscle cells, these other cells do not use their property of *rhythmicity*. Instead they follow the rate set by the sinu-atrial node using their property of *conduction*.

The heart beat spreads out from the sinu-atrial node through the atrial muscle [FIG. 11] and brings about contraction of both the atria. When these chambers contract they force their blood downwards through the atrioventricular valves into the ventricles [FIG. 3, p. 2].

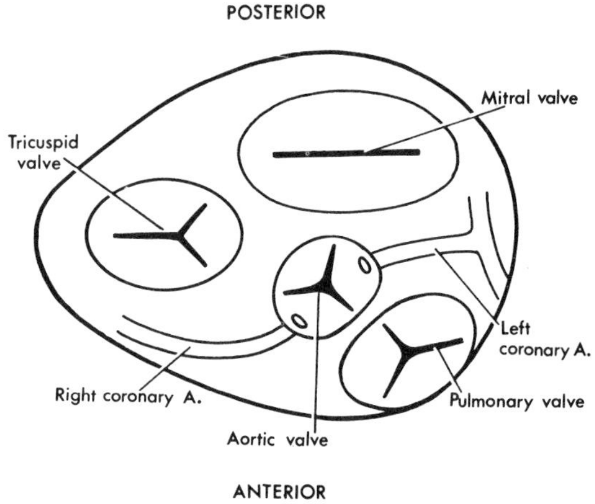

POSTERIOR

FIG. 12. The septum between the atria and the ventricles contains the four heart valves. The two coronary arteries originate from the aorta just distal to the aortic valve.

Between the atria and the ventricles there is a fibrous septum which contains the heart valves [FIG. 12]. This fibrous septum, not being made of cardiac muscle, will not conduct the cardiac impulse. As a result the contraction wave which has spread over the atria dies away instead of spreading directly to the ventricles. There is one pathway only between the atria and the ventricles. This starts at the **atrioventricular node** (A-V node), and runs down in the septum between the two ventricles, as the atrioventricular bundle (bundle of His). When the contraction wave, which is spreading over the atrial muscle, reaches the A-V node, it passes down this modified muscle tissue of the atrioventricular bundle, and reaches the ventricles via this route. It then spreads throughout the right and left ventricles and these ventricles contract.

When watching the heart beat at a thoracic operation the atria are seen to contract. There is a short pause, and the ventricles contract. After their contractions the atria and the ventricles go into a state of relaxation.

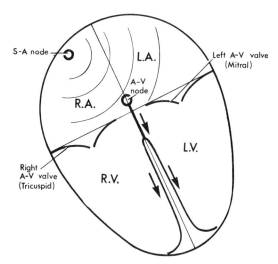

FIG. 11. Propagation of cardiac impulse through the heart. The heart beat starts at the sinu-atrial node (S-A node). It spreads through the musculature of the right atrium (R.A.) and left atrium (L.A.) to the atrioventricular node (A-V node). The contraction wave passes down the atrioventricular bundle (bundle of His) to reach the right ventricle (R.V.) and left ventricle (L.V.).

A heart beating in this normal manner with the heart beat originating at the S-A node is said to be in normal or **sinus rhythm.**

ELECTROCARDIOGRAM

This spread of the heart beat from the S-A node, first to the atria, then down the atrioventricular bundle, and finally to the ventricles, is associated with an electrical voltage change which can be recorded at distances remote from the heart. This electrical record is termed the **electrocardiogram** (ECG).

The voltage as recorded from the two arms (Lead I) is only one thousandth of a volt (one millivolt). It is

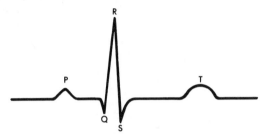

FIG. 13. The normal electrocardiogram. The waves are lettered alphabetically starting with P. The P wave corresponds to atrial systole. The QRST complex corresponds to ventricular systole.

amplified electronically and displayed using a pen recorder or a cathode ray oscilloscope. The apparatus is known as an **electrocardiograph.**

The recording shows fluctuations throughout the cardiac cycle which are known as waves and complexes [FIG. 13]. The P wave corresponds to atrial systole. It is followed by the **QRST** complex (a complex is a series of waves) which corresponds to ventricular systole.

The **QRST** complex recorded during ventricular systole is very small compared with the voltage which is actually produced by each cardiac muscle cell throughout the whole of ventricular systole. This is because the voltage produced by the right ventricle tends to cancel out the voltage produced by the left ventricle. In a similar way, if a flashlamp contains two torch batteries, and one of these batteries is inserted in a reverse manner so that the two batteries are back-to-back, the voltage supplied to the lamp-bulb is now so low that the lamp will not light. With the electrocardiogram, the voltage produced by one ventricle is back-to-back with that produced by the other, the over-all voltage recorded is very small.

Under normal circumstances, the only time when the cancellation is not complete is at the beginning of systole (**QRS** complex) and at the end of systole (**T** wave). The QRS complex is the result of the heart beat arriving at different parts of the ventricles at slightly different times. The **Q** wave is not always present. The **T** wave is due to some parts of the heart relaxing slightly before other parts.

If one ventricle does not produce its full voltage then

the cancellation will not be so complete. This will occur if the ventricle has an insufficient blood supply from the coronary arteries (**ischaemic heart disease, myocardial infarction** and **coronary thrombosis**).

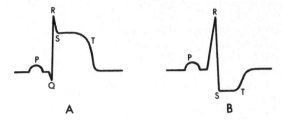

FIG. 14. The electrocardiogram of coronary thrombosis. A. Anterior thrombosis showing raised S-T segment. B. Posterior thrombosis showing depressed S-T segment. Recording using Lead I.

The failure of cancellation is seen most clearly in the part of the electrocardiograph tracing between the **S** wave and the **T** wave. In this part of the ECG, known as the **S-T segment,** the cancellation is usually so good that, although both ventricles are contracting and ejecting blood at this time, no voltage is recorded (isoelectric region).

Following a coronary thrombosis, this region of the electro-cardiogram will no longer be at zero voltage. The direction of displacement will enable the site of the thrombosis to be determined. FIGURE 14 shows the elevated and depressed **S-T segments** in Lead I of the electrocardiogram in two cases of coronary thrombosis. With an anterior thrombosis there is a raised **S-T** segment [FIG. 14(A)] and with a posterior thrombosis there is a depressed S-T segment [FIG. 14(B)].

Heart Block

The conduction pathway through the heart just described is of great importance in explaining many clinical disorders of conduction of the heart.

If a patient has damage to the atrioventricular bundle, then the heart beat, which originates at the S-A node, will spread to the atria and to the A-V node, but will not reach the ventricles. This condition is termed **heart block.** The ventricles stop beating, and the circulation of the blood ceases. The ventricles, however, are made of cardiac muscle, and, as such, have an inherent property of rhythmicity. In many cases the ventricles start beating, but at their own rate which is usually very much slower than that of the S-A node. A typical idioventricular rate is 30 beats per minute compared with the normal ventricular rate of 70 beats per minute which is the rate of the S-A node.

A ventricular rate of 30 beats/minute is usually insufficient to maintain an adequate circulation unless the person is lying flat. A faster ventricular rate is needed so that he can walk about, and take exercise. To speed up the heart a cardiac pacemaker is employed.

Cardiac Pacemakers

Cardiac muscle can be made to beat by electrical stimulation, and this fact is used in the treatment of a patient with a heart block. The problem is how to connect an electrical stimulator to the heart with the minimum of operative interference. Usually an electrode in the form of a catheter is passed under X-ray control through a superficial vein in the neck, down the superior vena cava into the right atrium and through the tricuspid valve into the right ventricle [FIG. 15]. The electrode is then in contact with the inside of the right ventricle. The external end is connected to a cardiac pacemaker and a neutral electrode is applied to the chest wall for the return pathway. An electric shock can then be sent to the heart at a rate of 60 or 70 times a minute. Each shock originates a beat which spreads to the two ventricles and causes ventricular contractions.

With a fixed rate pacemaker, the atria play no part in circulating the blood, since they will be contracting at a different rate. Nor will the ventricular rate increase in exercise although the S-A node and atrial rates may. In a more elaborate form of pacemaker the natural

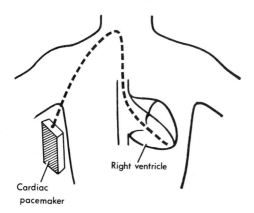

FIG. 15. Cardiac pacemaker. A catheter containing an electrode wire is passed into a neck vein, down the superior vena cava into the right atrium, and through the tricuspid valve into the right ventricle. This allows the ventricle to be stimulated electrically 70 times a minute. The return electrical path is via a neutral electrode in the axilla.

S-A node beat is used to trigger the pacemaker. In intermittent heart block, a demand pacemaker is used. This only produces a shock if the heart does not produce its own beat and no **QRS** complex appears.

Bundle Branch Block

If the heart block is restricted to one half of the atrioventricular bundle [FIG. 16], the condition is termed **bundle branch block**. The contraction wave can still spread from the A-V node to one ventricle and this ventricle will contract first. The contraction wave will then spread by the process of conduction to the other ventricle which will contract a short time later. This

delay in the spread of the contraction wave to the second ventricle shows itself in the long duration of the **QRS** part of the electrocardiogram [FIG. 16].

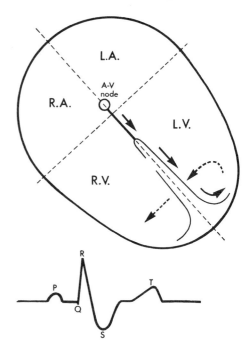

FIG. 16. Right bundle branch block. If the right branch of the atrioventricular bundle is interrupted, the contraction wave will reach the left ventricle directly, and only then will spread to the right ventricle. The tracing (*lower*) shows the electrocardiogram. The QRS complex occupies a longer period of time than normal because the isoelectric phase (ST) is not reached until the contraction wave has reached the right ventricle. R.A. = right atrium, L.A. = left atrium, R.V. = right ventricle, L.V. = left ventricle. A-V node = atrioventricular node.

P-R Interval

The interval on the electrocardiogram between the **P** wave and **QRS** complex (**P-R** interval) represents the time taken for the contraction wave to spread from the S-A node to the ventricles. This time is normally between 0·12 and 0·21 seconds.

In a disease such as rheumatic fever there is impairment of conduction in the atrioventricular bundle. The electrocardiogram then shows a prolonged **P-R** interval. A long **P-R** interval may be a forerunner of a complete heart block when there is no longer any constant time relationship between the **P** waves and the **QRST** complexes. Thus with complete heart block, the ventricular rate may be 30 beats a minute whilst the S-A node and atrial rate is 70 beats a minute. The electrocardiogram will then show 70 **P** waves every minute, but only 30 **QRST** complexes.

ECTOPIC BEATS

Occasionally another part of the heart, other than the S-A node, originates a heart beat. It will be remembered that each cardiac muscle cell has the property of rhythmicity, but normally the other cells do not use their power of beating and passively follow the beat originated by the S-A node.

Atrial Extrasystoles

If another cell in the atrium originates a heart beat it is termed an **ectopic focus** and this beat is termed an **atrial extrasystole**. It shows itself on the electrocardiogram as an abnormal **P** wave which is different in shape

to contract again until an interval of time has elapsed. This is termed the **refractory period.** When an additional beat originates from an ectopic focus, the next normal beat, which has originated from the S-A node, may arrive at the ventricles during the refractory period. If so, it will not bring about a ventricular contraction. There will be a ventricular pause until the next beat from the S-A node. This may be seen in FIGURE 19 where, following a succession of normal heart beats, a ventricular extrasystole appears and is followed by electrical silence until the next beat but one appears.

A person in whom these extrasystoles are appearing seldom notices the additional beat but thinks that the heart has missed a beat because of the long pause that

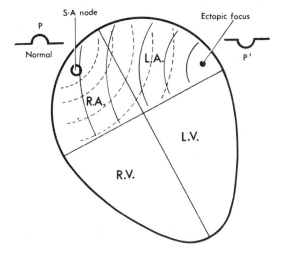

FIG. 17. Atrial extrasystole. An ectopic focus in the atria will give rise to an abnormal P wave (P¹). The QRST complex will be normal. R.A. = right atrium, R.V. = right ventricle, L.A. = left atrium, L.V. = left ventricle, S-A node = sinu-atrial node.

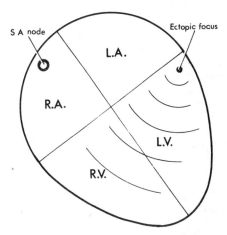

FIG. 18. Ventricular extrasystole. An ectopic focus in the ventricles gives rise to a ventricular extrasystole. This will not be preceded by a P wave. R.A. = right atrium, R.V. = right ventricle, L.A. = left atrium, L.V. = left ventricle, S-A node = sinu-atrial node.

and may be inverted. This inversion occurs when the wave of contraction is spreading through the atrial muscle in the opposite direction to normal [FIG. 17].

When this abnormal contraction wave reaches the A-V node, it will pass down the A-V bundle to the ventricles and produce a normal **QRST** complex. Isolated atrial extrasystoles are of little clinical significance.

Ventricular Extrasystoles

These are much commoner and are of much greater importance. If an ectopic beat originates from a ventricular cardiac muscle cell a contraction wave will spread over the ventricular muscle, and bring about a ventricular contraction [FIG. 18]. But this ventricular contraction will not have been preceded by an atrial contraction. The electrocardiogram associated with this ventricular extrasystole will show an abnormally large and abnormally shaped **QRST** complex [FIG. 19, p. 12], but there will have been no preceding **P** wave.

Once the ventricles have contracted, they are unable

follows. This is termed a **dropped beat** although in fact no beat has been lost, there has been instead a **premature beat**. Just as one may be unaware of a clock ticking, but notices immediately the ticking stops, so one may be unaware of the heart beating but notices the apparent missed beat.

Ventricular extrasystoles are a comparatively common occurrence, especially during states of emotional excitement, and probably occur at some time or other in most people. It is only when extrasystoles are occurring frequently that they become clinically significant, since they may then be a forerunner of ventricular fibrillation.

A patient who is receiving digitalis may produce an extrasystole after each normal beat. This is termed 'coupled beats' and is an indication that the dose of digitalis should be reduced.

Adrenaline increases the likelihood of extrasystoles, and an over-dose of this drug will cause ventricular fibrillation.

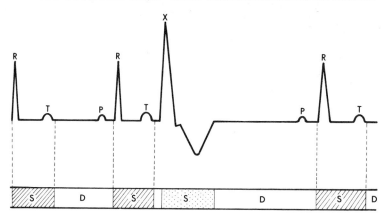

FIG. 19. Electrocardiogram of ventricular extrasystole (X). The extrasystole is abnormal in shape and is not preceded by a P wave. The next P wave which occurs during the extrasystole does not bring about a ventricular contraction and there is a delay until the P wave after that produces a ventricular systole. S = systole; D = diastole. Cross-hatched shading indicates normal systole; dotted shading indicates extrasystole; unshaded area between represents a very short diastole.

Ventricular Fibrillation

In **ventricular fibrillation** many ectopic foci are present in the ventricles and, as a result, there is no organized contraction of the ventricles as a whole. Circulation of the blood ceases, and, unless immediate resuscitation measures are taken [see cardiac resuscitation, p. 62], death ensues.

It appears that extrasystoles which occur during the T wave of the previous beat, are most likely to bring on ventricular fibrillation, since the cardiac muscle has an increased excitability during this phase.

Atrial Fibrillation

If a number of ectopic foci are present in the atria, the condition is termed **atrial fibrillation.** This is not so serious as ventricular fibrillation since atrial contraction is not essential for life [see functions of atria, p. 6]. The appearance of the fibrillating atria has been likened to that of a bag of wriggling worms. The electrocardiogram shows an absence of any obvious P waves, but the A-V node receives a very large number of contraction waves from different parts of the atria and transmits these to the ventricles. The ventricles however are unable to respond to all of these, and, as a result, the ventricular systoles occur at frequent but irregular intervals. A patient with atrial fibrillation has a pulse that is *irregularly irregular* in its timing, that is, the heart rate is fluctuating with no set pattern and it is impossible to predict when the next beat is coming. There may be long intervals, there may be a series of very rapid beats and these two states may alternate.

ELECTRICAL AXIS OF THE HEART

The beating heart produces its electrical voltage in the general direction from the centre of the chest [point **O**, FIGURE 20] diagonally to the left. This direction, shown by the arrow in FIGURE 20, is termed the **electrical axis of the heart.**

This electrical voltage (which gives rise to the electrocardiogram) is a **vector** because it has both **magnitude** and **direction.** [Other examples of vectors include the *wind* which has a magnitude (wind velocity) and a direction (the direction in which it is blowing), and the *radio signal* from the local radio or television station

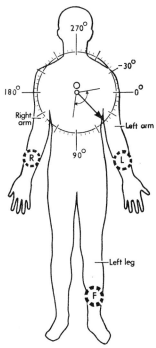

FIG. 20. Recording the electrocardiogram. Electrodes are applied to the right arm (R), the left arm (L), the left leg (F) and to the chest.
The electrocardiogram is produced by the cardiac vector which lies in a direction from the centre of the chest towards the left side of the body. It is shown by an arrow.

which has a magnitude (field strength) and a direction. In order to receive a powerful television signal, a television aerial must be pointed directly towards the television station. Similarly if a transistor radio is rotated, it will give its maximum volume when its internal aerial is in line with the signal from the radio station. If it is now rotated through a right angle the volume will decrease to a minimum.]

Since the electrical voltage produced by the heart is a vector—**the cardiac vector**—the electrocardiogram will have a large amplitude when it is recorded in the

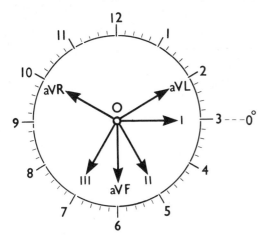

FIG. 21. The six standard limb leads, I, II, III and aVR, aVL and aVF record the component of the cardiac vector in different directions. They have been shown here with reference to an imaginary clock-face superimposed upon the chest. O corresponds to the centre of the chest.

same direction as this cardiac vector. If it is recorded at a right angle to the cardiac vector the amplitude will be small. At more than a right angle the electrocardiogram will be inverted.

The direction of the cardiac vector in the frontal plane may be found by recording the electrocardiogram in a number of different directions and then seeing which direction shows the largest amplitude. The six standard leads employed are termed **I, II, III, aVR, aVL,** and **aVF.**

If one imagines a clock-face superimposed on the front of the chest, with its centre at point **O,** and that the clock only has an hour hand, then this hand will represent the direction in which the electrocardiogram is recorded. The hand pointing to 3 o'clock will give the direction of Lead **I,** to 5 o'clock Lead **II,** to 7 o'clock Lead **III** [FIG. 21]. Lead **aVL** will be represented by the hand pointing to 2 o'clock, **aVF** to 6 o'clock and **aVR** to 10 o'clock. The cardiac vector, itself, usually lies between 2.30 and 6.30 using this analogy with a mean direction of about 4.30. The direction of Lead **II** will then be closest in direction to this cardiac vector [compare FIGS. 20 and 21] and as a result the ECG recorded in Lead **II** is usually larger than that recorded in any other lead [see FIG. 23].

In practice the clock-face nomenclature is replaced by a 0–360 degrees circular protractor scale with the 0 degrees at 3 o'clock [FIG. 20]. Using degrees, Lead **I** is horizontal to the left at 0 degrees.

$$\text{Lead } \mathbf{II} = 60°$$
$$\text{Lead } \mathbf{III} = 120°$$
$$\mathbf{aVF} = 90°$$
$$\mathbf{aVR} = 210°$$
$$\mathbf{aVL} = 330° \text{ (usually referred to as } -30°)$$

In **left ventricular hypertrophy** the cardiac vector is deflected to the left (**left axis deviation**) and lies in the general direction from the centre of the chest to the left shoulder (2 o'clock or $-30°$) [FIGS. 20 and 21]. Lead **aVL** will have the largest amplitude. Lead **I** will also be large and upright but Lead **III** will be inverted (more than a right angle).

In **right ventricular hypertrophy,** the cardiac vector is deflected to the right and lies in the general direction from the centre of the chest to the right side of the body (more than 6.30 on the clock-face or 100°) [FIGS. 20 and 21]. This is termed **right axis deviation.** Lead **I** will be inverted whereas Lead **III** will be large and upright.

Chest leads V_1 to V_6 may also be recorded. They enable the direction of the cardiac vector in the transverse plane to be determined.

A FURTHER CONSIDERATION OF LEADS EMPLOYED

Limb Electrodes

Electrodes are applied to the right arm (R), the left arm (L) and the left leg (F = foot). The right leg is not normally employed but an electrode may also be applied here to earth the subject. The electrodes usually take the form of a metal plate which is strapped to the wrist and ankle, with electrode jelly underneath to ensure a good electrical connexion with the skin. Alternatively a dry electrode with projections (like a nutmeg grater) may be employed.

Electrically the arms are an extension of the trunk and it is immaterial whether the electrodes are placed at the wrist, on the upper arm or at the shoulder.

Standard Limb Leads I, II, and III

All three limbs are connected to the electrocardiograph and the appropriate combination chosen by internal switching. The amplifier connexions made when recording the three standard limbs leads [FIG. 20] are:

LEAD **I**: right arm (R) and left arm (L)
LEAD **II**: right arm (R) and left leg (F)
LEAD **III**: left arm (L) and left leg (F).

The electrocardiogram is a record of the voltage difference between two points (joined by dotted lines, FIG. 22). Two connexions have to be made to the electrocardiogram amplifier. When Lead **I** is recorded one connexion is from each of the two arms [FIG. 22 (*upper*)]. With Leads **II** and **III** one arm and one leg are connected to the amplifier. These leads are termed **bipolar leads.** Typical records are shown in FIGURE 23 (*upper*).

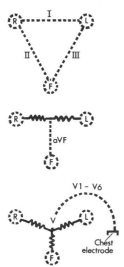

FIG. 22. The electrical connexions made in the electrocardiogram apparatus when recording Leads I, II, III (upper), aVF (middle) and V₁ to V₆ (lower). The position of the amplifier is indicated by the dotted lines.

Augmented Limb Leads aVR, aVL and aVF and Chest Leads

An electrocardiogram can be recorded using one search electrode only provided that a neutral point can be found to provide the second amplifier connexion. This is termed a **unipolar lead.** When the search electrode is applied to the chest it is termed a **chest electrode.**

It has been shown that if all three limbs are connected electrically to a common point **V** through electrical resistors [FIG. 22 (*lower*)] then the voltage at **V** does not fluctuate throughout the cardiac cycle. This provides the required neutral point for the second connexion to the amplifier. It may be considered as equivalent to an electrode at point **O** [FIG. 20] at the centre of the chest.

If the search or chest electrode is applied to the left leg, the voltage between **V** and the left leg (**F**) will be recorded (Lead **VF**). However, the electrical resistor between the **V** point and the left leg (**F**) serves only to shunt the amplifier and reduce its gain. If it is removed, the record will be larger and this is termed the **augmented VF** lead or Lead **aVF** for short. The two remaining electrical resistors between (**R**) and (**L**) are shown in FIGURE 22 (*middle*).

When Lead **aVL** is recorded, the resistor between **V** and **L** is removed, leaving that between **R** and **V** and that between **F** and **V**.

With Lead **aVR**, the resistor between **V** and **R** is removed, leaving those between **F** and **V**, and **L** and **V**.

Chest Leads V₁ to V₆ [FIG. 22 (*lower*)]

When the chest leads V₁ to V₆ are recorded, all three resistors are present and the **V** point is connected to one input of the amplifier. The chest electrode, which usually takes the form of a suction cup which adheres to the chest, is applied in turn to the following positions:

V₁ 4th right interspace at sternal border
V₂ 4th left interspace at sternal border
V₃ mid-way between V₂ and V₄
V₄ 5th left interspace and mid-clavicular line
V₅ same level as V₄ and anterior axillary line
V₆ same level as V₄ and mid-axillary line

V₁ and V₂ ECG tracings usually show large **S** waves whilst V₅ and V₆ show large **R** waves [FIG. 23 (*lower*)]. In V₃ and V₄ the **R** and **S** waves are of approximately equal amplitude.

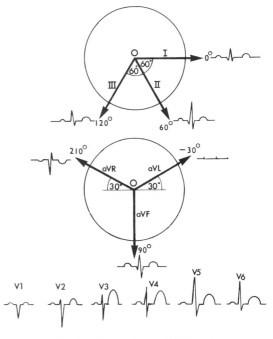

FIG. 23. The 12 standard ECG leads.

Upper figure. Leads I, II and III. Lead I records the component of the cardiac vector in the horizontal plane (0°). Lead II records it at 60° to this plane. Lead III records it at a further 60° (=120° to the horizontal plane).

Middle figure. Leads aVF, aVR and aVL. aVF records it at an angle of 90° (vertically downwards). aVR at an angle of 210°. aVL at an angle of 330° (−30°). The lead which has the direction closest to the cardiac vector will show the largest amplitude. This enables the direction of the cardiac vector to be determined.

Lower figure. Leads V₁, V₂, V₃, V₄, V₅, V₆. The chest electrode is applied in turn to the six chest positions (see text) and Leads V₁ to V₆ recorded.

4. BLOOD PRESSURE

Units Employed

By tradition pressures in the body are measured in millimetres of mercury (mm. Hg), and not dynes per square centimetre, pascals or pounds per square inch.

The pressures are measured with reference to the atmospheric (barometric) pressure and not with reference to a vacuum. In other words only the amount by which the pressure exceeds the barometric pressure (P_B) is recorded.

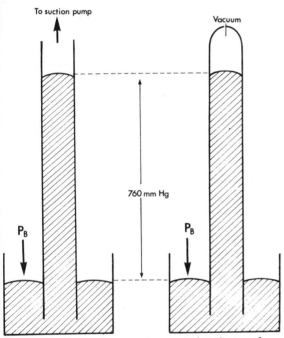

Fig. 24. If a suction pump is connected to the top of a metre glass tube immersed in mercury, the mercury is pushed up the tube by the barometric pressure, P_B, acting on the mercury in the reservoir. It rises to a height of 760 mm. Hg. The tube may be sealed off and the suction pump removed leaving a vacuum at the top of the tube. This acts as a barometer. P_B is the barometric pressure.

Barometric Pressure

If a vacuum pump is attached to the top of a long tube immersed in mercury, the mercury rises up the tube. No matter how perfect the vacuum pump, the mercury only rises to a height of about 760 mm. Hg and then stops rising [Fig. 24, *left*]. The tube may be sealed off and the vacuum pump removed, leaving a vacuum at the top [Fig. 24, *right*].

The mercury has not been sucked up the tube by the vacuum, but has been pushed up by the pressure of the atmospheric air (P_B) acting on the mercury in the reservoir. The vacuum prevents this pressure acting on the mercury in the tube. Such a device acts as a baro-

meter, and the height of the mercury column is termed barometric pressure. The height of the mercury column fluctuates with atmospheric conditions. At sea-level the barometric pressure is usually in the range of 730–780 mm. Hg with a mean of 760 mm. Hg (= one atmosphere).

The barometric pressure decreases with increase in altitude and halves every 18,000 feet. Thus it will be 380mm. Hg at 18,000 feet above sea-level and only 190 mm. Hg at 36,000 feet—the height at which many jet air-liners fly. The cabin of such a plane is pressurized to maintain an environmental pressure close to that at sea-level.

Blood Pressure

Such a device, as shown in Fig. 24, could be used as a mercury manometer to record blood pressure by converting the mercury reservoir into a sealed chamber with an inlet tube which would allow the pressure to be applied via air or saline to the mercury [Fig. 25].

The apparatus with a vacuum at the top would, however, be unwieldy since when recording arterial blood pressure the mercury would rise to a height of 880 mm.

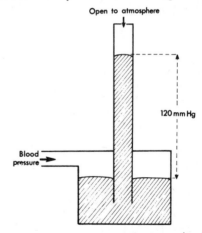

Fig. 25. Mercury manometer for measuring blood pressure. The blood pressure is applied to the mercury in the reservoir via air or fluid. The tube is open to the atmosphere at the top. The pressure of 120 mm. Hg with this apparatus means 120 mm. Hg above atmospheric pressure (or 880 mm. Hg absolute).

(120 mm. more than the barometric pressure). A tube in excess of 3 feet long would be required. Furthermore the reading obtained would depend on atmospheric conditions. It would fluctuate from day to day with changes in the barometric pressure. A systolic blood pressure of 120 mm. Hg might raise the mercury to a height of 890 mm. on a fine day but only 860 mm. on a wet day.

In order to measure only the amount by which the blood pressure exceeds the barometric pressure, the differential manometer principle is used with the mercury column open to the atmosphere at the top [FIG. 25]. This allows the air pressure to be applied to the top of the mercury column. A much shorter tube can now be employed. To prevent mercury leakage during storage, in the clinical version of this apparatus (sphygmomanometer) a seal at the top of the tube which is mercury-tight but not air-tight is employed.

0 mm. Hg on the sphygmomanometer scale represents atmospheric pressure (760 mm. Hg with reference to a vacuum). 120 mm. Hg means 120 mm. Hg above the atmospheric pressure (760 + 120 = 880 mm. Hg with reference to a vacuum).

In the thorax and veins of the skull the pressure may be less than the atmospheric pressure (say 755 mm. Hg with reference to a vacuum). Such a sub-atmospheric pressure is often referred to as − 5 mm. Hg. The minus sign indicates that the pressure is 5 mm. Hg below the atmospheric pressure.

As an alternative to the mercury manometer to measure pressure, an anaeroid manometer or an electronic manometer may be employed. The pressure displaces a diaphragm and the movement of this diaphragm is converted into the movement of a needle round a dial (anaeroid) or an electrical change by means of a transducer such as a strain gauge (electronic). An electronic manometer, connected to a catheter, may be used to record pressure changes in the heart and blood vessels. These manometers are calibrated using a mercury manometer.

VENTRICULAR PRESSURES

The pressure in the left ventricle during diastole is at atmospheric pressure (0 mm. Hg). As the left ventricle

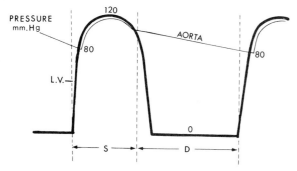

FIG. 26. Pressure changes in the left ventricle and aorta. Left ventricular pressure fluctuates between 0 and 120 whereas aortic pressure fluctuates between 120 and 80 mm. Hg. S = systole, D = diastole, L.V. = left ventricle.

contracts the pressure rises, and it reaches a peak of 120 mm. Hg during systole. It then falls off slightly and by the end of systole has fallen to about 110 mm. of mercury. With the onset of diastole the ventricular pressure falls rapidly down to 0 mm. Hg [FIG. 26, *heavy curve*].

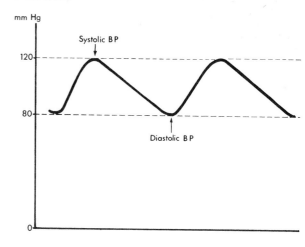

FIG. 27. Fluctuations of blood pressure in an artery. The maximum pressure reached is termed the systolic blood pressure (systolic B.P.). The lowest pressure is termed the diastolic blood pressure (diastolic B.P.).

The aorta is in continuity with the left ventricle during systole, and its pressure too reaches 120 mm. Hg [FIG. 26, *light curve*], but as the left ventricular pressure starts to fall the aortic valve shuts. The elastic recoil of the aorta and arteries near to the heart maintains a blood pressure during diastole and allows blood to continue to flow on to the tissues, even though there is no output from the heart during this phase of the cardiac cycle. The pressure in the aorta has only fallen to 80 by the time the next systole occurs and the pressure is increased once again to 120.

The maximum pressure in the aorta and large arteries is termed the **systolic blood pressure,** the minimum pressure is termed the **diastolic blood pressure** [FIG. 27]. It will be seen that the pressure in the aorta is fluctuating between a peak value of 120 mm. Hg and a minimum value of 80 mm. Hg. This is written as:

Blood Pressure = 120/80 mm. Hg.

On the right side of the heart [FIG. 28] the same

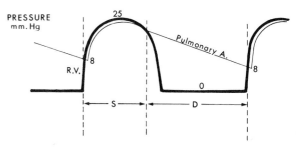

FIG. 28. Pressure changes in the right ventricle and pulmonary artery. Right ventricular pressure fluctuates between 0 and 25 mm. Hg. Pulmonary artery pressure fluctuates between 25 and 8 mm. Hg. S = systole, D = diastole, R.V. = right ventricle.

quantity of blood is pumped out, but under a much lower pressure. The right ventricular pressure in diastole is 0 mm. Hg, and this increases to only 25 mm. Hg during systole.

The pulmonary artery pressure will follow the right ventricular pressure up to a peak of 25 mm. Hg, and the pulmonary valve will shut as the right ventricular pressure starts to fall. The pulmonary artery is not so elastic as the aorta and its pressure has fallen to 8 mm. Hg by the time the next systole has occurred [FIG. 28]. The pulmonary artery pressure is therefore fluctuating between a peak of 25 mm. Hg and a trough of 8 mm. Hg. Thus the pulmonary artery pressure is 25/8 mm. Hg.

The pressures throughout the circulation are summarized in FIGURE 29.

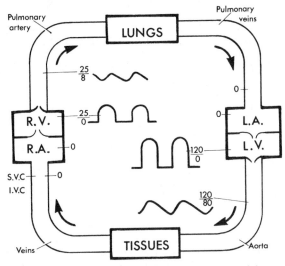

FIG. 29. Summary of pressures in different parts of the circulation. Where two pressures are shown, the upper pressure is the pressure during systole and the lower pressure is that during diastole.

ARTERIAL BLOOD PRESSURE

The expression 'blood pressure', if unqualified, refers to the pressure in the aorta and large arteries (arterial blood pressure). Pressures in other parts of the circulation are prefixed by the vessel referred to, e.g. pulmonary artery blood pressure, capillary blood pressure, etc.

The organs of the body require a **blood flow.** The arterial blood pressure is needed to 'push' the blood through the arterioles, capillaries and veins in order to achieve this blood flow. This driving force from behind is termed in Latin *vis a tergo*. Provided that the blood flow is adequate, the arterial blood pressure which achieves this flow is of secondary importance. It has been seen that in the pulmonary circulation quite a low pulmonary artery blood pressure (16 mm. Hg mean) is adequate to maintain a 5,000 ml./minute lung blood flow, whereas in the systemic circulation a high aortic pressure (100 mm. Hg mean) is required.

Some races of South-east Asia maintain this systemic blood flow with a mean pressure of 70 mm. Hg and thus have lower blood pressures than are common in the Western world.

But there are two further factors to be taken into account concerning blood pressure. Firstly, in the sitting or standing position the brain is at a higher level than the heart, and a blood pressure is needed to pump the blood up-hill from the heart to the brain. Too low a blood pressure will lead to an inadequate cerebral blood flow. This pressure is no longer required when lying flat. Thus a person with a very low blood pressure may remain conscious when lying down, but will lose consciousness if moved to the sitting or the standing posture.

The second reason for a blood pressure is to enable the kidneys to make urine [p. 75].

FACTORS DETERMINING BLOOD PRESSURE

In order to have a blood pressure there must be a cardiac output and a resistance to blood flow in the systemic circulation. This resistance is termed the **peripheral resistance.**

Blood Pressure = Cardiac Output × Peripheral Resistance

The factors affecting cardiac output are discussed in CHAPTER 5 but at rest it is comparatively constant and as a result the blood pressure will be determined principally by the peripheral resistance. This resistance to

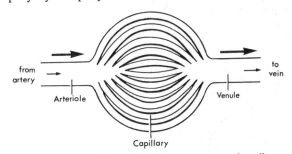

FIG. 30. Arteriole supplies a large number of capillaries. Although each capillary has a smaller diameter than that of the arteriole, the arteriole offers a higher resistance to blood flow than the capillary network.

blood flow lies mainly in the small arteries of the body which are termed **arterioles.** It is the small diameter vessels that offer the greatest resistance to blood flow. $\left(\text{Resistance is proportional to } \dfrac{1}{\text{diameter}^4}\right)$. The capillaries are even smaller vessels than the arterioles, but, although each individual capillary will offer a higher resistance than an arteriole, there is a large number of capillaries in parallel supplied by each arteriole [FIG. 30]. As a result there is a large number of alternative pathways for the blood to take in its passage from an arteriole through to the veins, and because of this the capillary network does not offer such a resistance to blood flow as the arteriole supplying it.

Viscosity of Blood

The resistance offered by an arteriole of a given size depends on the viscosity of the blood. Blood is a sticky viscous fluid which offers two to three times more resistance than plain water or saline. The viscosity of blood depends partly on the plasma and partly on the number of red cells present.

The viscosity of the blood is usually a constant, but it will be reduced if large quantities of saline are given. Plasma substitutes, such as dextran, are viscous fluids. A reduction in the circulating red cells (anaemia) has little effect on the viscosity but it will be increased in polycythaemia (high red cell count). A low blood viscosity will be associated with a low blood pressure and a high viscosity with a high blood pressure.

CENTRAL CONTROL OF BLOOD VESSELS

The arterioles have smooth muscle in their coat and this smooth muscle is circularly arranged round the blood vessels. When this muscle contracts it makes the

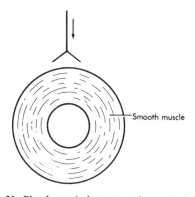

FIG. 31. Blood vessels have smooth muscle in their wall. This muscle contracts whenever impulses reach the blood vessel via the sympathetic nervous system and reduces the size of the lumen. This state is termed vasoconstriction.

blood vessel smaller [FIG. 31]. The arterioles of the body would be in a state of large size, **vasodilatation,** were it not for the fact that they have a nerve supply from the sympathetic nervous system, which is acting upon them and producing a state of **vasoconstriction** as a result of the contraction of the smooth muscle. This sympathetic activity, or **sympathetic tone** as it is called, originates from a collection of cells in the medulla termed the **vasomotor centre** (VMC).

Three neurones are involved in transmitting the information from the VMC to the arteriole [FIG. 32]. The first runs from the medulla down the spinal cord in the lateral columns of white matter to the lateral horn cells of grey matter which are found in the thoracic and upper lumbar segments.

The second neurone preganglionic fibre leaves the spinal cord via the ventral nerve roots as the white

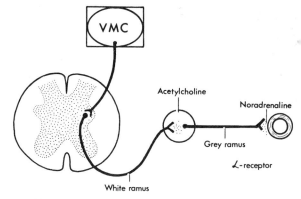

FIG. 32. Nerve pathway from vasomotor centre to blood vessels. VMC = vasomotor centre.

ramus and runs to the sympathetic trunk. Here it synapses with the third neurone or postganglionic fibre (grey ramus) which runs to the blood vessel, usually in the trunk of an anatomical mixed nerve of the body [CHAPTER 19].

The chemical transmitter at the postganglionic termination is **noradrenaline.**

The vasomotor centre is keeping the arterioles of the body in a state of partial vasoconstriction. If the activity of the vasomotor centre **increases,** the arterioles become smaller, that is, there is more **vasoconstriction.**

Should the vasomotor centre activity **decrease** then the blood vessels will revert to their initial large size; this state is termed **vasodilatation.**

Under normal conditions an increase in vasomotor centre activity which causes **vasoconstriction,** will bring about an **increase** in blood pressure, whilst a decrease in vasomotor centre activity will lead to **vasodilatation** and a **lowering** of blood pressure [FIG. 33].

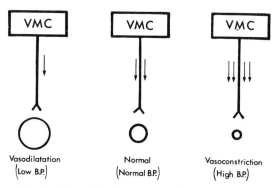

FIG. 33. Size of blood vessels depends on sympathetic activity which is regulated by the vasomotor centre. VMC = vasomotor centre.

Baroreceptors

The blood pressure is maintained at a constant level by the **baroreceptors.** These are sensory receptors found in the wall of blood vessels in the region of the carotid sinus, the brachiocephalic trunk and the aortic arch [FIG. 34]. These receptors are sensitive to blood pressure and send information concerning the blood pressure to the vasomotor centre in terms of coded nerve impulses.

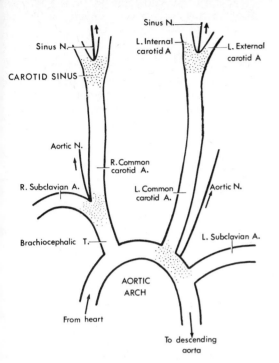

FIG. 34. Baroreceptors in the circulation. The areas in the aortic arch and brachiocephalic trunk send baroreceptor activity up the aortic nerve (branch of vagus). Baroreceptor activity from the carotid sinus passes along the sinus nerve, a branch of the glosso-pharyngeal nerve (IX).

The nerve activity that passes along a typical barore-ceptor neurone at three levels of blood pressure may be seen in FIGURE 35. It will be seen from FIGURE 35 that the activity fluctuates during the cardiac cycle. This is because the blood pressure varies during the cardiac cycle. It will also be seen that the nerve activity in-creases with increase in the blood pressure.

This baroreceptor activity acts as a 'brake' on the vasomotor centre activity. The baroreceptor nerve activity is said to inhibit the vasomotor centre. Thus the greater the baroreceptor activity, the smaller will be the sympathetic tone to the arterioles, and the larger will be the vessels [FIG. 36].

Such a mechanism provides *negative feedback* which will tend to minimize any change in blood pressure. If for any reason the blood pressure starts to fall, as for example, after a blood donor has given a pint of blood, the activity of the baroreceptor nerves will decrease. Fewer inhibitory impulses will now pass to the vaso-motor centre. The inhibitory brake will be removed and the vasomotor centre activity will increase. As a result there will be increased vasoconstriction. This will in-crease the peripheral resistance, which will prevent more than a very slight fall in blood pressure.

Conversely, if there is any tendency for the blood pressure to rise, the baroreceptor activity will increase. This will further inhibit the vasomotor centre activity. The sympathetic tone to the arterioles will decrease and the arterioles will dilate. This vasodilatation will lower the peripheral resistance and limit the rise in blood pressure.

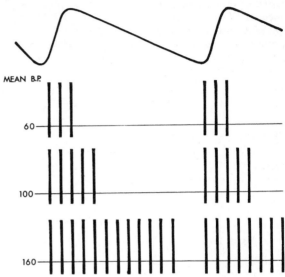

FIG. 35. Typical activity in single baroreceptor neurone at three different mean blood pressures. It will be noted that baroreceptor activity occurs during systole and early diastole and that there is a pause in late diastole. Top tracing, fluctuations in blood pressure.

When the blood pressure rises the baroreceptor activity increases. This increase in baroreceptor acti-vity:

(a) inhibits the vasomotor centre and causes vaso-dilatation;

it also acts on:

(b) the cardiac centre in the medulla and slows the heart;

(c) the respiratory centre and depresses breathing.

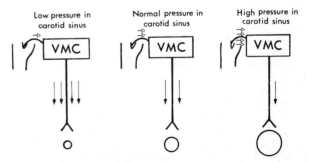

FIG. 36. Baroreceptor activity inhibits sympathetic vasoconstrictor tone to blood vessels. Baroreceptor activity tends to minimize any change in blood pressure and the baroreceptor nerves are termed 'buffer nerves'.

Other Factors Affecting the Vasomotor Centre

In addition to the baroreceptors there are other factors which modify the vasomotor centre activity, and, by altering the sympathetic activity to arterioles, affect the blood pressure [FIG. 37].

In a conscious person the most important additional factor is the effect of the **higher centres.** By higher centres is meant the higher parts of the central nervous system and includes the regions of the cerebral cortex where conscious thought occurs. Emotional excitement and stress are accompanied by a stimulation of the vasomotor centre by higher centres which brings about increased vasoconstriction, increased peripheral resist-

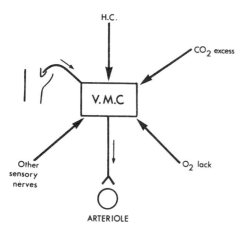

FIG. 37. Principal factors affecting the activity of the vasomotor centre. VMC = vasomotor centre, HC = higher centres.

ance, and hence an increase in blood pressure. Thus the blood pressure recorded under the stress of a clinical examination may be higher than it is when there is no stress.

On the other hand, there are occasions when higher centres decrease the activity of the vasomotor centre so that vasodilatation occurs, the peripheral resistance falls and so does the blood pressure. Should the blood pressure fall to a low level, then the subject will faint. This is what occurs when a patient faints at the sight of blood, or a nurse faints on her first visit to the operating theatre. It is due to the reduction of the vasoconstrictor tone to the arterioles and also, as will be seen later, to a reduction in the vasoconstrictor tone to veins.

Sufficient carbon dioxide must be maintained in the blood in order to allow the vasomotor centre to function properly. Over-breathing is harmful because it washes the carbon dioxide out of the blood, and, one of the effects of this, is to cause a fall of blood pressure, due to a decrease in the vasomotor centre activity [see p. 57].

A shortage of oxygen, on the other hand, will stimulate the vasomotor centre both directly and via the chemoreceptors [p. 55], and in the early stages of oxygen lack (anoxia) the blood pressure rises due to an increase

in vasomotor activity. Chemoreceptors also stimulate the vasomotor centre following a haemorrhage.

Many other nerves in the body affect the vasomotor centre activity. In general, moderate pain will stimulate the vasomotor centre and cause an increase in blood pressure, whereas very severe pain may inhibit the vasomotor centre and lead to fainting.

LOCAL CONTROL OF BLOOD VESSELS

The blood flow to the organs of the body is related to their needs, and local control mechanisms exist to enable the flow to be increased when required.

MUSCLE BLOOD FLOW

The muscles need an increased blood flow during exercise and this is brought about by the vasodilatation of the arterioles as a result of the production of **metabolites.** These metabolites are waste-products produced by the muscle metabolism, and consist of carbon dioxide, potassium ions and other substances. They act on the arterioles and override the sympathetic vasoconstrictor tone, so that the arterioles dilate.

Reactive Hyperaemia

If the circulation to a limb is occluded, say by inflating a blood pressure cuff to above the systolic blood pressure (200 mm. Hg) for five minutes, the metabolites will accumulate in the limb. When the circulation is restored, their vasodilating action can be seen. The skin becomes warm and flushed. Measurements of muscle and skin blood flow show that the blood flow has increased markedly.

Ischaemic Pain

Muscular activity, when there is insufficient blood flow (ischaemia) to remove the metabolites, leads to the ischaemic pain. This pain may be produced by placing the arm high above the head and clenching and unclenching the fist rapidly. Due to the height above the heart, the blood flow to the limb is now inadequate. If, as a control, a similar action is carried out with the other arm by the side, ischaemic pain will develop in a short time in the raised arm without any pain being experienced in the arm by the side.

Intermittent Claudication

The discomfort which appears in the leg muscles during walking, due to ischaemia of these muscles, is termed **intermittent claudication.**

It is not present at rest, appears after walking a certain distance, and disappears in a few minutes when the walking stops.

This accumulation of metabolites is due to a narrowing or obstruction to the femoral or popliteal arteries. In severe cases, the popliteal, posterior tibial or dorsalis pedis pulses may be absent [FIGS. 44, 45, pp. 23, 24]. An arterial graft may be undertaken to restore the continuity of the blood vessel.

HEART MUSCLE BLOOD FLOW

The heart requires an increased blood supply when it is producing a large cardiac output, such as in exercise. This is brought about by dilatation of the coronary vessels. These coronary vessels dilate when the surrounding cardiac muscle is short of oxygen. The heart thus has a local control mechanism for regulating its own blood supply. If the cardiac muscle cells are not receiving enough oxygen for their needs, the coronary blood vessels dilate, and thus increase the blood supply to the cardiac muscle.

Angina Pectoris. The pain associated with ischaemia of cardiac muscle (myocardial ischaemia), due to a partial obstruction to the coronary blood vessels, is termed angina pectoris. The pain is brought on by exercise. The pain is not localized to the heart, but is a referred pain which takes the form of a tight constricting pain round the chest which may extend down the inside of the arms.

Nitrites, and certain nitrates, dilate coronary blood vessels. Glyceryl trinitrate tablets, for example, allowed to dissolve slowly under the tongue, or swallowed in the form of slow-release capsules, are used to relieve acute attacks of myocardial ischaemia. They are also used to forestall an attack of angina pectoris by being taken before going for a walk. (These compounds unfortunately also dilate cerebral blood vessels and may give rise to 'nitrite headaches'.)

With a complete blockage of the coronary blood vessels (coronary thrombosis) the pain persists even at rest.

SALIVARY GLAND BLOOD FLOW

The salivary glands provide an example of an organ which increases its blood supply by nervous activity. The salivary glands have a parasympathetic nerve supply which is vasodilator, and increases the blood supply to the glands when more saliva is being produced. This is important since saliva is made from blood.

SKIN BLOOD FLOW

The skin is the only organ of the body whose blood supply is not related to the organ's metabolic needs. The blood supply to the skin is regulated in the interests of the regulation of temperature by the body as a whole. When the body is too **hot,** the skin blood vessels are dilated. Blood is sent to the skin, so that heat may be lost. When the body is too **cold,** the skin blood vessels are constricted. Blood is kept away from the skin, and this conserves heat and maintains the body temperature. This alteration in skin blood flow is brought about partly by the direct action of heat on the skin blood vessels (**heat dilates blood vessels; cold constricts them**), and partly by the action of the temperature regulating centre, situated in the hypothalamus, which alters the sympathetic tone to the skin blood vessels. With **heat, the sympathetic tone is reduced,** and the blood vessels dilated. With **cold, the sympathetic tone is increased,** and the blood vessels constrict. The skin blood vessels

are also sensitive to mechanical and chemical stimulation, and their response is termed skin reactions.

SKIN REACTIONS

White Line

If the blunt end of a pin is stroked across the skin, then after a delay of about 15 seconds, a white line appears over the area stroked. This is due to capillary constriction, and the white line is due to the fact that the blood has been squeezed out of the underlying capillaries.

Triple Response

If the sharp end of the pin is used, then, instead of a white line, a red line appears. With this firmer pressure, the capillaries dilate, and it is this capillary dilatation that shows as the red line [FIG. 38].

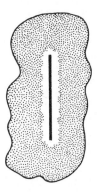

FIG. 38. Triple response. The red line and wheal are surrounded by the flare.

Surrounding the red line is a red flare, and this is due to arteriolar dilatation. The dilatation of the arterioles only occurs if the sensory nerves are intact, and is due to what is termed an **axon reflex.** The sensory nerves from the skin, which run in via the dorsal nerve roots of the spinal cord, send branches to the blood vessels in the vicinity of the sensory receptors [FIG. 39].

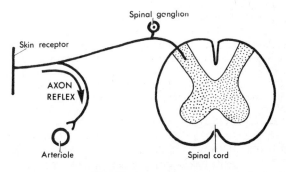

FIG. 39. Axon reflexes.

The sensory receptors in the skin send nerve impulses not only into the spinal cord but also round via an axon reflex to the arterioles. It is these impulses that bring about the dilatation of the arterioles. The area of the **flare** has an arteriolar distribution, and shows an irregular margin. The area also becomes raised because of the increased production of tissue fluid as a result of an increase in the permeability of the capillaries. The elevated area is termed a **wheal**. The three events which occur, namely, red line, the flare and the wheal are termed the **triple response** [FIG. 38].

The skin produces the same triple response, no matter how it is injured. Scratching with a pin, acids, alkalis, heat, cold electric currents will all produce the same response. It is thought that all these agents produce skin damage, and this is associated with the release of histamine. An injection of histamine, itself, into the skin, will produce a typical triple response.

Histamine is also released in the skin as a result of an antigen-antibody reaction. Thus when a person becomes allergic to a form of food, such as strawberries or lobsters, the protein from this food acts as an antigen, and the body makes a substance (antibody) which will destroy this antigen. The antigen-antibody reaction releases histamine, and it is the release of the histamine that produces the skin rash associated with these allergies. The released histamine also causes bronchospasm [p. 58].

Antihistamine drugs are substances which have the property of opposing the action of histamine, and they will therefore reduce the triple response.

Excessive Vasodilatation Leading to Blood Pressure Fall

The local mechanisms over-ride the central vasoconstrictor tone and allow local vasodilatation to occur. In order to maintain the same blood pressure, more vasoconstriction will usually be needed in the parts of the vascular tone still under vasomotor centre control. If this vasoconstriction is not possible the blood pressure will fall and the person will faint. In severe cases, the circulation may fail completely. Fainting may occur, for example, when exercise is taken in a very hot environment, or in anaphylactic shock which is associated with an excessive release of histamine.

ADRENAL MEDULLA HORMONES

The adrenal medulla releases a mixture of adrenaline and noradrenaline (catecholamines) as hormones. Noradrenaline has a generalized vasoconstriction action, whilst adrenaline constricts skin blood vessels but dilates muscle blood vessels. They both increase the force of contraction of the heart [p. 29].

ANGIOTENSIN

Angiotensin is a small peptide (9 amino acids) which is formed when renin is released by the kidneys [p. 44].

Angiotensinogen + renin → angiotensin
(*plasma protein*)

Angiotensin circulates in the blood and causes vasoconstriction [see hypertension, p. 26].

Summary of Factors Influencing the Size of an Arteriole

The principal factors which determine the size of an arteriole are summarized in FIGURE 40.

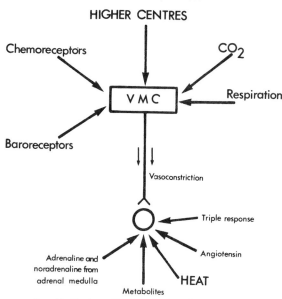

FIG. 40. Factors affecting the size of an arteriole.

THE PULSE

FIGURES 41–45 show the sites in the body where the arterial pulsations can be most readily palpated and often seen.

When using the pulse to determine the heart rate the object is to determine the number of complete cardiac cycles there are in one minute. The timing should therefore start with the first pulse, and this first pulse should be counted as **0**. The next should be counted as **1**, the next as **2**, and so on.

A ruler or tape-measure used to measure distance starts at **0**, not **1**, and the same consideration applies to heart rate counting.

The radial pulse is the pulse most frequently employed to determine the heart rate. It is important to remember that it is the blood pressure changes in the radial artery that are being felt when the pulse is taken. The rapid rise in pressure from 80 mm. Hg to 120 mm. Hg with systole is transmitted rapidly through the arterial tree at a rate of about six metres per second and this pressure change takes about $\frac{1}{10}$ second to reach the wrist.

Care should be taken to differentiate between blood pressure and blood flow. The blood ejected from the heart with each beat flows much more slowly and takes many seconds to reach the wrist and does not arrive until several beats later.

Although the presence of a pulse confirms that the main blood vessel channels are patent between the heart and the site of taking a pulse, the fact that a pulse is

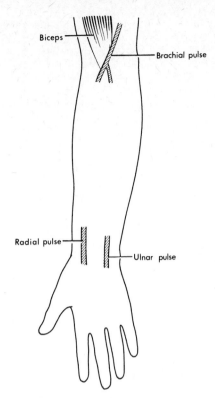

FIG. 41. Pulses palpable in the forearm.
To determine the pulse rate, counting should start at
0 not 1.

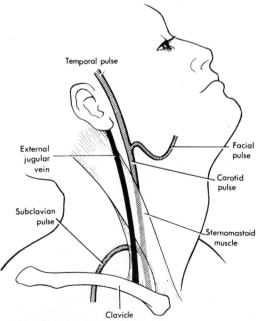

FIG. 42. Pulses in the head and neck region. Venous
pulsation in the external jugular vein is usually only
seen when the subject is lying down. This pulsation
cannot be palpated.

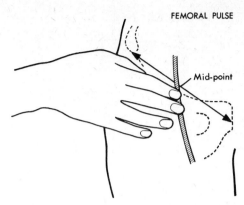

FIG. 43. Femoral pulse. This is found midway
between the anterior superior iliac spine and the sym-
physis pubis.

absent does not necessarily mean that there is no blood
flow in the artery. It is the fluctuation of blood pressure
in the artery between the systolic pressure (120 mm. Hg)
and the diastolic pressure (80 mm. Hg) that gives rise to
the pulse. If for any reason there is no difference between these pressures, and the pressure is a mean pressure of 100 mm. Hg, there would still be an adequate
blood flow, but no pulse would be detected.

It is the presence of a resistance to blood flow
proximal to the point of measuring that eliminates the
pressure difference between systole and diastole. Thus,
the presence of the arteriolar resistance removes the
pressure fluctuations from the capillaries. Similarly an
obstruction to the arterial tree will eliminate or modify
the pulse. In the case of a complete obstruction, the
blood will flow through anastomotic channels.

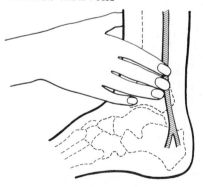

FIG. 44. Posterior tibial pulse. This is found behind
the medial malleolus.

When lying down, a **jugular venous pulse** can usually
be seen at the side of the neck [FIG. 42] due to small
pressure changes which are transmitted from the right
atrium. It shows three maxima (a, c, v) per cardiac
cycle. This venous pulse is too feeble to be palpated.

DORSALIS PEDIS PULSE

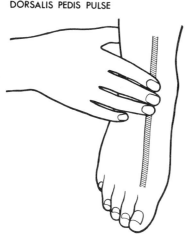

FIG. 45. Dorsalis pedis pulse. This is midway between the two malleoli.

Coarctation of the Aorta

An obstruction to the arch of the aorta at the site of the ductus arteriosus is termed coarctation of the aorta. The blood reaches the lower limbs via anastomoses and pulsations in the enlarged intercostal arteries may sometimes be felt when placing one's hands on the back to lift the patient.

The pulses in the lower limbs may be reduced or absent and the femoral blood pressure will be lower than the brachial blood pressure.

MEASUREMENT OF BLOOD PRESSURE

The standard method of taking a patient's blood pressure is to use the technique developed by Korotkov in 1905. An inflatable arm cuff is applied round the upper arm, snugly but not too tightly, leaving a clearance of two inches between the lower end of the cuff and the cubital fossa, at the elbow. This cuff is inflated using a small hand pump and the pressure in the cuff is measured using a mercury manometer [FIG. 46]. This apparatus is termed a **sphygmomanometer.**

The brachial artery pulse is located in the cubital fossa at the elbow by palpation. It lies on the medial side of the biceps tendon, and the arterial pulsations can often be seen if the arm is fully extended.

Note that a stethoscope cannot be used to locate the brachial artery since the flow in this artery is laminar, [see p. 4], and no sounds will be heard until the cuff has been inflated.

The radial pulse is next palpated at the wrist, and, whilst the fingers of one hand are on this pulse, the other hand is used to inflate the cuff to a pressure well above the point at which the radial pulse disappears. This ensures that the cuff has been inflated to well above any silent interval if present [FIG. 46].

The stethoscope is then placed over the brachial artery, and the pressure in the cuff slowly lowered. In order to maintain a steady fall of pressure, the release

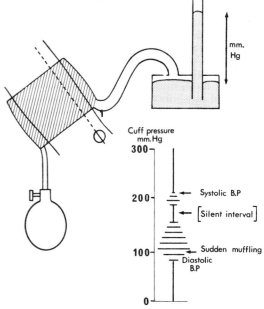

FIG. 46. Recording brachial artery blood pressure using a sphygmomanometer. The cuff is inflated to above the systolic pressure (as determined by the absence of the arterial pulse). A stethoscope is applied over the brachial artery at the elbow and the pressure lowered steadily. The sounds heard (Korotkov sounds) are represented by the horizontal lines in the lower part of the figure. The intensity of the sound is indicated by the width of the line. The point at which tapping sounds are first heard is the systolic blood pressure. The point at which these sounds become suddenly muffled is the diastolic blood pressure.

Occasionally it is found that, as the cuff pressure is reduced, the 'tapping sounds' disappear over a range between the systolic and diastolic pressures. This is known as the silent interval. The silent interval is not normally present.

valve must be opened more and more as the pressure falls.

As the pressure descends, no sounds are heard until the systolic blood pressure is reached, when tapping sounds, corresponding to the heart rate, are heard in the stethoscope. The pressure at which these sounds first appear is noted. This gives the systolic blood pressure. As the cuff pressure continues to be reduced, the sounds become louder, and louder, but at the diastolic pressure they suddenly change their quality and become muffled. A little lower down they finally disappear never to reappear again. The point at which the sounds become muffled is taken as the diastolic pressure.

Silent Interval. Occasionally when recording the blood pressure of a hypertensive patient a 'silent interval' is found [FIG. 46]. As the cuff pressure is reduced from 300 mm. Hg, the tapping sounds may start at say 220 mm. Hg, indicating a high systolic blood pressure. Around 180 mm. Hg the sounds disappear and reappear again around 150 mm. Hg giving a 'silent interval' between these two pressures. As the cuff pressure falls still further the sounds become suddenly muffled at 100 mm. Hg and at 85 mm. Hg disappear, never to reappear again. This patient's blood pressure is thus 220 mm. Hg systolic and 100 mm. Hg diastolic.

Although it occurs comparatively rarely, the silent interval offers 'a trap for the unwary'. It is most likely to catch out those who routinely pump the cuff up to 160 mm. Hg or so, instead of inflating it until the radial pulse has disappeared, as described above.

It will be seen from Fig. 46 that the whole of the upper range of Korotkov sounds may be lost and the systolic blood pressure reported as 150 mm. Hg instead of its true value of 220 mm. Hg.

Alternatively, the systolic pressure might be correct but the diastolic pressure recorded incorrectly as 175 mm. Hg.

When the sounds have disappeared shortly below the recorded 'diastolic pressure', it is essential to continue the fall in cuff pressure to ensure that the sounds do not reappear again. This confirms that one is not on the top part of the Korotkov sound range.

When using the stethoscope, it is essential to insert it into the ears the right-way round. The ear pieces should be in a downward and forward direction when viewed from above. Care should be taken not to touch the rubber tubing otherwise extraneous sounds will be produced. Since the Korotkov sounds are very faint, it is not possible to take blood pressure accurately in noisy surroundings.

It is important not to keep the cuff **inflated longer than necessary** and to allow the cuff pressure to **fall to zero** after each determination.

With most people, the blood pressure fluctuates by as much as 10 mm. Hg or so with breathing. It is therefore pointless to record the blood pressure of such an individual with an accuracy of one millimetre of mercury (i.e. 117/82) unless the phase of respiration at the instant of taking each of the two readings is also recorded! For most purposes a reading to the nearest 5 mm. Hg (or even 10 mm. Hg) is sufficient, and, in any case, is all that is justified by the method.

Femoral Blood Pressure

A similar technique may be employed to take the femoral (popliteal) artery blood pressure. Owing to the increased diameter of the thigh, a wider blood pressure cuff has to be used, and this is referred to as a thigh-cuff. The subject lies face downwards, and the thigh-cuff is applied to the thigh [Fig. 47].

An identical method to that employed in the arm is then used to measure the femoral blood pressure. The posterior tibial artery is palpated as it runs behind the medial malleolus (in place of the radial artery). The stethoscope is placed over the centre of the popliteal fossa at the back of the knee instead of the cubital fossa at the elbow [Fig. 47]. The femoral blood pressure is determined using the Korotkov sounds as in the case of the brachial pressure determination. These sounds start

when the cuff pressure has been reduced to a value equal to the femoral systolic blood pressure, and they become muffled when the diastolic pressure is reached.

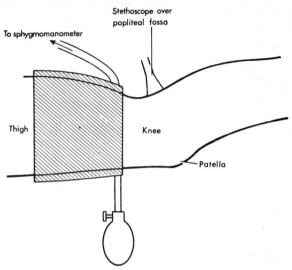

FIG. 47. Determining femoral blood pressure using a sphygmomanometer. A thigh cuff, which is wider than an arm cuff, is applied to the thigh with the subject lying face downwards. The cuff is inflated to above the femoral systolic blood pressure (as determined by the disappearance of the posterior tibial or dorsalis pedis pulse [FIGS. 44 and 45]. A stethoscope is applied to the middle of the popliteal fossa and the pressure on the thigh cuff lowered steadily. The same Korotkov sounds are heard as with the brachial cuff on the arm.

Recording Blood Pressure in a Pulseless Patient

If the patient is pulseless, no Korotkov sounds will be heard. In such a case the blood pressure may be determined in the following manner. The subject's arm is held vertically so as to drain the veins and capillaries of blood. The cuff is then raised rapidly to 200 mm. Hg or higher to arrest the blood flow. The arm is then lowered. It is now pale in colour.

The cuff pressure is lowered in 10 mm. Hg steps, waiting after each pressure lowering to see if the colour changes, marking a return of blood to the arm. The cuff pressure at which these colour changes occur, and at the same time the veins on the back of the hands begin to swell, is noted. This pressure will be just below the systolic blood pressure, or, if there is no pulsation and the systolic and diastolic blood pressures are the same, just below the mean blood pressure.

GRAVITY AND BLOOD PRESSURE

Owing to gravity the blood pressure increases by 10 mm. Hg for every 12 cm. below the heart. Above the heart it decreases by the same amount. Thus the systolic blood pressure will be 210 mm. Hg in the feet but only 90 mm. Hg in the brain in the upright posture.

When lying down these two pressures will be the same (120 mm. Hg).

HIGH BLOOD PRESSURE [HYPERTENSION]

If the blood pressure is too high the condition is termed **hypertension.** The exact level at which the blood pressure ceases to be normal and enters the hypertensive range has been the subject of much discussion since it is **blood flow** that is important and not blood pressure. It has been suggested that 'hypertension exists when the blood pressure is higher than the person looking after the case thinks it ought to be'!

The normal systolic blood pressure is 120 mm. Hg at rest, but it may rise to 160 mm. Hg or higher in emotional excitement. The normal diastolic pressure is 80 mm. Hg at rest. It too rises in emotional excitement but to a smaller extent. A diastolic pressure of over 110 mm. Hg at rest is unusual and probably warrants the label of hypertension. In severe hypertension no doubt exists. Systolic blood pressures in excess of 300 mm. Hg have been recorded.

Hypertension places a strain on the left ventricle of the heart, since this chamber has to pump out blood at the increased pressure. It leads ultimately to heart failure [p. 44].

The high blood pressure in hypertension may cause blood vessels to burst. The lenticulostriate arteries in the brain are particularly vulnerable, and their rupture (cerebral vascular accident) leads to a cerebral haemorrhage which interrupts the motor fibres from the motor cortex to the spinal cord. These fibres cross to the other side in the medulla. The result is paralysis in the other side of the body (hemiplegia).

(A thrombosis or spasm of these vessels, not necessarily associated with hypertension, will produce similar symptoms.)

Hypertension is the result of too high a cardiac output or too high a peripheral resistance. In the majority of cases, the cardiac output appears to be normal, and the methods of treating the condition are usually aimed at reducing the peripheral resistance.

FIGURE 48 shows the principal sites where steps may be taken to reduce the blood pressure.

The effect of higher centres on the vasomotor centre may be reduced by the use of sedatives and tranquillizers.

In hypertension the baroreceptors have been reset to the new level. As a result the baroreceptor activity is producing insufficient inhibition of the vasomotor centre to keep the blood pressure down. This inhibition may be increased by stimulating the baroreceptor nerves electrically using implanted electrodes. It should be remembered, however, that the sinus and aortic nerves contain both baroreceptor and chemoreceptor fibres and that stimulation of the chemoreceptor nerve fibres will increase the blood pressure still further!

The vasomotor centre activity reaches the blood vessels via the sympathetic nervous system. The sympathetic pathways may be interrupted by a sympathectomy. Alternatively ganglionic blocking agents, such as

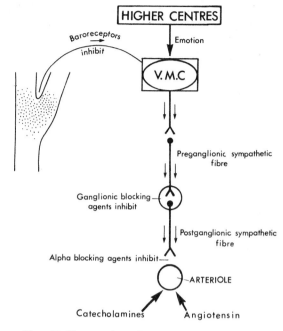

FIG. 48. Hypertension. Factors affecting arteriolar tone. A lowering of the peripheral resistance and hence a lowering of blood pressure may be brought about by a number of methods which inhibit the release of noradrenaline at the postganglionic sympathetic termination.

hexamethonium, may be used to prevent the transmission of nerve impulses from the preganglionic sympathetic fibres to the postganglionic sympathetic fibres in the sympathetic ganglion [p. 113].

Noradrenaline is released as the chemical transmitter at the postganglionic sympathetic termination. Drugs may be employed which inhibit the synthesis of noradrenaline, block its release and action, or cause its rapid destruction at this site.

The cause of the majority of cases of hypertension is unknown (essential hypertension). Occasionally the treatment of the cause cures the condition. For example, an ischaemic kidney will release **renin** which acts on the plasma proteins and forms the peptide **angiotensin.** Angiotensin is a powerful vasoconstricting agent and if it is formed in sufficient amounts hypertension will develop. Following removal of this diseased kidney, leaving the other normal kidney, the blood pressure returns to normal.

A phaeochromocytoma tumour of the adrenal medulla endocrine gland [p. 104] produces excessive amounts of the hormones adrenaline and noradrenaline which cause hypertension by their action on the heart and blood vessels. Removal of the tumour cures the hypertension.

Cushing's disease [p. 105] due to excessive production of corticoids by the adrenal cortex is associated with hypertension due to the increased blood volume.

PRESSURE DROP THROUGH THE CIRCULATION

The blood pressure in the aorta has a maximum value of 120 mm. Hg, and a minimum value of 80 mm. Hg. It has a mean value of about 100 mm. Hg. The blood flows from the arteries to the arterioles and through the capillaries to the veins, because there is a progressive fall in blood pressure, and blood flows from a region of high pressure to a region of lower pressure. There is only a very small drop in pressure as the blood flows through the arteries, but the pressure falls markedly as the blood flows through the arterioles. The pressure falls from 100 mm. Hg at the arterial end to 32 mm. Hg at the capillary end of the arteriole. There is a further fall of pressure to 12 mm. Hg as the blood flows through the capillaries, and a further fall from 12 mm. Hg down to 0 mm. Hg as the blood returns along the veins.

The large fall in pressure which is found in the arterioles, is due to the fact that the arterioles are offering a resistance to flow. Thus, if a large number of people are trying to leave a hall quickly through a small door, there will be a high pressure in the hall, but once a person has passed through the resistance, namely the door, the pressure will disappear. There is thus high pressure before a resistance, and a low pressure after the resistance.

In the circulation the main peripheral resistance lies in the arterioles, and there is thus a high pressure before the arterioles and a low pressure after the arterioles.

FIGURE 49 shows the fall of pressure as the blood flows from the arteries through the arterioles and the capillaries to the veins. Just as a stream runs down hill, so the blood flows from a region of high pressure to a region of low pressure.

Due to the elasticity of the aorta and large arteries, and to the resistance offered by the arterioles, the capillaries are supplied with blood continuously. The elastic recoil of the arteries maintains the flow during diastole when no blood is leaving the ventricles. If an artery is cut the blood spurts out. Each spurt is the result of the blood pressure increasing from 80 mm. Hg up to 120 mm. Hg with each ventricular systole. A cut capillary does not spurt; the blood flows out steadily. The same applies to a cut vein. This is because the pressures in the capillaries and the veins do not fluctuate with each heart beat.

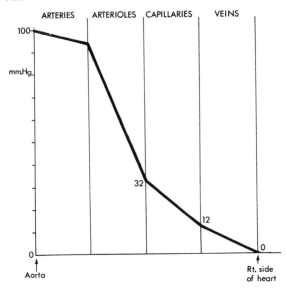

FIG. 49. The mean blood pressure falls as the blood flows through the arteries to the arterioles, capillaries and veins.

Summary:

Blood pressure is determined by the cardiac output [CHAPTER 5] and peripheral resistance.

$$B.P. = C.O. \times P.R.$$

1. The peripheral resistance depends principally on the size of the arterioles. It also depends upon the viscosity of the blood.
2. Arterioles are kept in a state of tonic vasoconstriction by the vasomotor centre (VMC) in the medulla acting via the sympathetic nervous system (sympathetic tone). This tone maintains the peripheral resistance and hence the blood pressure.
3. Baroreceptors moderate the activity of the VMC and 'buffer' changes in blood pressure.
4. Higher centres acting via the VMC may increase blood pressure, as in emotional excitement, or may reduce blood pressure and cause fainting, as in severe pain or shock.
5. Local mechanisms can affect the size of arterioles in an organ and may over-ride the central VMC control.

5. CARDIAC OUTPUT

The amount of blood pumped out by each ventricle per minute is termed the **cardiac output.** It depends on the product of two factors: the heart rate and the stroke volume.

CARDIAC OUTPUT = HEART RATE × STROKE VOLUME

It should be remembered that it is always the **heart rate** multiplied by the **stroke volume** that is important in determining the rate at which blood circulates in the body and that the cardiac output cannot be measured using the heart rate alone. However, if the heart rate increases, then provided the stroke volume is unchanged, the cardiac output will increase.

The normal cardiac output at rest is 5,000 ml./min. Since we are dealing with a circulation, this will not only be the rate at which blood leaves the left ventricle. It will be the rate at which blood arrives via the veins at the right side of the heart [FIG. 50]. 5,000 ml./min.

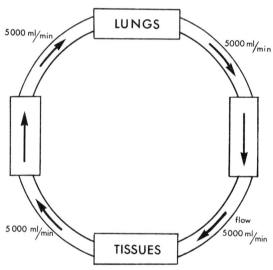

FIG. 50. The rate of blood flow is the same throughout the circulation. The blood passing through the lungs each minute is the same as that pumped out by each ventricle. Hence cardiac output = lung blood flow.

will be the rate at which blood leaves the right ventricle, the rate of blood flow through the lungs and the rate at which blood returns to the left side of the heart via the pulmonary veins.

The factors which modify heart rate and stroke volume will be considered in turn.

HEART RATE

The heart rate is determined by the rate at which the S-A node beats, and, as has been seen, this S-A node has its own inherent rhythmicity. The natural rate of

beating of the S-A node is, however, much higher than the rate normally found which is about 70 beats/minute. This is because of the action of the vagus nerve which supplies the heart. Its activity acts as a brake on the rate at which the S-A node beats. This inhibitory activity is termed **vagal tone.** 'Tone' is the name given to nervous activity which is maintained continuously.

Since the vagus is acting as a brake on the heart rate, the greater the vagal tone the slower the heart, and the smaller the vagal tone the faster the heart rate. Thus, when during sleep the vagal tone increases, the heart rate slows. On awakening the vagal tone decreases and the heart rate speeds up.

The vagus (tenth cranial nerve) is part of the parasympathetic nervous system. The action of the vagus on the heart is blocked by the parasympathetic blocking agent *atropine* [p. 117]. The heart rate of a patient will thus speed up when this drug is administered.

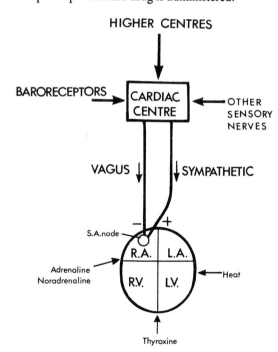

FIG. 51. Factors affecting heart rate.

This vagal tone originates in the collection of cells in the medulla known as the **cardiac centre** [FIG. 51].

The heart is also under the influence of the sympathetic nervous system which acts on the S-A node and causes the S-A node to beat at a faster rate. The sympathetic nervous system also acts on the ventricular muscle and brings about a more powerful contraction of

28

the ventricles. Sympathetic activity is increased in states of emotional excitement, and during exercise, and both of these states are associated with a fast heart rate.

The sympathetic activity brings about its effect by the release of **noradrenaline** at the nerve ending. This substance, and a closely related substance, **adrenaline,** is released as a hormone from the **adrenal medulla** gland, and hormones from this gland will augment the activity of the sympathetic nervous system in speeding up the heart in states of emotional excitement.

The action of the sympathetic nervous system on the heart is blocked by drugs such as *propranolol* and *practolol* which are sympathetic beta-blocking agents [p. 116]. When these drugs have been administered the heart rate no longer increases with emotional excitement.

Other factors which affect the rate at which the S-A node beats by a direct action on this node include body temperature and circulating thyroxine from the thyroid gland. Any increase in body temperature will increase the heart rate. The heart beats faster than normal in a fever. The heart beats slower if the body is cooled as in hypothermia.

Increase in activity of the thyroid gland which results in an increase in the circulating hormone thyroxine would increase the rate at which the S-A node beats, and patients with over-activity of the thyroid gland will therefore have fast heart rates [see p. 102].

The cardiac centre, itself, in the medulla, is under the influence of many factors. The most important in conscious man is probably the **higher centres.** Emotional stress increases heart rate. This is seen in competitors waiting for a race to start; their heart rate often increases before the gun goes off, and before they have taken a single step.

Increases in heart rate in times of stress up to 200 beats per minute have been demonstrated in patients during ward rounds, in airline pilots coming in to land jet planes at large airports, and in motor racing drivers waiting for the race to start and during the race itself. These heart rate increases are brought about by the effect of higher centres on the cardiac centre.

The baroreceptors act on the cardiac centre and their action is to slow the heart when the blood pressure goes up, and to speed up the heart when the blood pressure falls. Thus the fast heart rate which is seen in a patient following a haemorrhage (the pulse is rapid and thready) is due to the fact that there is now less baroreceptor activity, and as a result the heart rate speeds up.

Sensory nerves have a variable effect on heart rate; slight pain will give an increase in heart rate, whereas very severe pain may bring about reduction in heart rate.

An increase in venous return to the heart leads to an increase in heart rate especially when the heart rate is slow. This is known as the **Bainbridge Reflex.**

STROKE VOLUME

Stroke volume is defined as **the amount of blood pumped out by each ventricle per beat.** It is about 70 ml./beat at rest, but may increase to 150 ml./beat with exercise. The stroke volume is equal to the amount of blood in the ventricle when systole starts, less the amount of blood in the ventricle when systole ends.

Capacity of the Circulation and Blood Volume

The heart has very little power of sucking blood into itself (*vis a fronte*) and the volume of blood that is in the heart at the beginning of systole depends on how much blood is left over after the circulation has been filled. The blood volume is 5,000 ml. The capacity of the blood vessels must therefore be less than this, so that the difference between the two will, at any time, be in the heart and available for pumping out. This is represented diagrammatically in FIGURE 52 by a pair of scales. On the one side is the total amount of blood in the body. This volume is greater than the volume on the other side which is the volume of blood in the blood vessels. This volume is the capacity of the circulation. The difference between the blood volume and the capacity of the circulation is indicated by the pointer at the top. This pointer is deflected from the zero position and is at the 70 ml. position to represent the normal state. The blood volume is 5,000 ml., the capacity of the circulation is 4,860 ml. The remaining 140 ml. are in the heart, 70 ml. in the left ventricle and 70 ml. in the right ventricle.

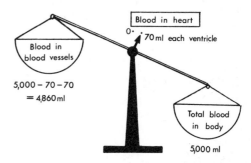

FIG. 52. Blood available in heart to be pumped out at next systole is the difference between the total blood in the body and the capacity of the blood vessels. Should the blood volume fall or the capacity increase, the output of the heart will be reduced.

Should the blood in blood vessels, that is, capacity of circulation, increase, then the scales would become more equally balanced, and as a result there would be less blood left over to fill the heart, and the stroke volume would fall. Thus the stroke volume is determined by the difference between the capacity of the circulation and the blood volume.

Venomotor Tone. It is the veins that hold most of the blood in the body, and it is the capacity of the veins that is very important in maintaining the stroke volume. The veins are thin-walled vessels, but they have a muscular coat, and this muscular coat is supplied by sympathetic nerve fibres. The continuous activity that is running from the VMC to the veins is termed **venomotor tone,** and we rely on the venomotor tone in order to keep the capacity of our veins down to a sufficiently low level so that we have a reasonable stroke volume.

In the upright posture gravity is pulling the blood down to the lower parts of the body, down to the feet, and is tending to distend the veins. This venous distension must be overcome by the venomotor tone. If the venomotor tone is unable to counteract the effects of gravity, the veins will distend, the capacity of the circulation will become greater than the blood volume, and the stroke volume will fall to zero. As a result the cardiac output will also fall to zero and the patient will lose consciousness.

After a stay in bed of a few weeks the venomotor tone is lost, and if such a patient is suddenly taken out of bed and stood upright, he will faint due to the fact that there is very little blood returning to the heart.

A similar state of affairs may occur in a patient who is given a sympathetic or ganglionic blocking agent to reduce the sympathetic activity. In excessive doses there will be no effective control of the venomotor tone. Such a patient will be conscious when lying flat, but will faint when standing upright. This is called **postural hypotension.**

A venous distension can be prevented by external pressure applied to the veins. This may take the form of bandaging the legs and lower abdomen. A more effective mechanism is to apply a pressure suit. Ideally the pressure should be adjusted so as to just counteract the effect of the gravity.

Haemorrhage and Cardiac Output

The effect of haemorrhage is to reduce the blood volume, and referring back to FIGURE 52 it will be seen that the effect will be to reduce the blood volume side of the balance. As a result the stroke volume will be reduced and will fall towards zero. The stroke volume may be increased again by giving a blood transfusion. If this is not possible, the stroke volume can be improved by reducing the capacity of the circulation, namely by lying the patient flat, and raising the legs, or applying a pressure suit to the limbs and lower part of the abdomen.

Starling's Law of the Heart

The ventricular muscle has a very important property. If more blood returns to the heart during the filling phase, this blood will distend the ventricles. At the next systole the ventricles will produce a more powerful contraction, and will pump out an increased volume of blood. The ventricles are thus able to transfer blood from their venous side to their arterial side, and the amount of blood they pump out will usually be determined by the amount of blood they receive. This fact

was first shown by Starling, and is known as Starling's Law of the Heart. It is often defined as saying that **the force of contraction of the ventricles depends on the initial length of the ventricular muscle fibres.**

Starling's law of the heart is very important in ensuring that the two sides of the heart have the same output per minute. It will be realized that if this were not so, the circulation of blood would rapidly cease. If, for example, the left ventricle pumped out more blood than the right ventricle, blood would very rapidly leave the lungs completely, and all the blood would be in the tissues. Conversely, if the right ventricle pumped out more blood per minute than the left ventricle, all the blood would be transferred, over the course of time, from the systemic circulation to the lungs. This does not occur because, should the right ventricle produce a greater output than the left ventricle, more blood would return to the left ventricle than had been pumped out, the left ventricle would be distended with this increased venous return, and it would give a more powerful contraction, and would increase its own stroke volume. Thus Starling's law of the heart maintains equality between the two sides of the heart in terms of cardiac output.

Starling's law of the heart accounts for the changes in the heart's activity following a haemorrhage. Since there is less blood in the ventricles at the end of diastole, the heart makes only a feeble contraction with the next systole giving a low stroke volume and a *weak pulse.* When a subsequent blood transfusion has increased the end-diastolic volume of blood in the ventricles, the heart makes a powerful contraction and pumps out this increased volume of blood, thus giving an increased stroke volume and a strong pulse.

Valves in Veins

The veins of the limbs have valves which allow the flow of blood only in one direction, that is, back towards the heart. These valves break up the column of blood and prevent distension of the veins due to gravity. They also act as a muscle pump which increases the return of blood to the heart.

This pump requires the *alternate contraction* and *relaxation* of the surrounding muscle. A sustained contraction or a continued relaxation will not operate this muscle pump.

With each contraction the blood in a segment of vein is squeezed towards the heart. With the next relaxation, this segment of vein fills from the distal end, and the alternate contraction and relaxation of the muscle will keep the blood moving back towards the heart [FIG. 53].

There is thus very much less venous distension when walking about than when standing still, and there is thus a very much better stroke volume and cardiac output when a person is walking than when standing.

It often happens that soldiers faint during a military parade. This is most likely to occur when they are

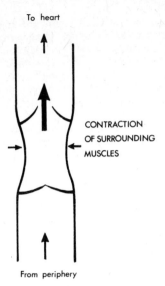

To heart

CONTRACTION
OF SURROUNDING
MUSCLES

From periphery

FIG. 53. Valves in veins and muscle pump.

standing still, and the muscle pump is not operating. As a result the veins are slowly distending under gravity, and this is increasing the capacity of the circulation and thus reducing the stroke volume.

Fainting Due to Low Cardiac Output

Fainting occurs when there is insufficient brain blood flow to maintain consciousness. In this case it will be due to a cardiac output which is inadequate to maintain the blood pressure necessary to oppose the effect of gravity and pump blood uphill from the heart up to the brain.

The act of fainting causes the person to fall to the ground, and to bring about an immediate improvement in cerebral blood flow.

A person who faints should be left in the horizontal position where, not only is the brain at the same level as the heart, but also the distending effect of gravity on the leg veins has been removed, until he himself recovers and sits up. Dragging such an unconscious person to his feet may cause brain damage due to cerebral ischaemia. For this reason fainting may have much more serious consequences if it occurs in a packed crowd so that the person is jammed in the vertical position, and does not fall to the ground.

As has already been seen, fainting also occurs when there is insufficient peripheral resistance, due to a loss of arteriolar tone.

Shock

A patient in shock is anxious, pale, sweating and has a low blood pressure and *rapid thready* pulse. He has a low cardiac output.

Shock may result from trauma, burns, sepsis, acute heart failure or a massive fluid loss due to diarrhoea.

The treatment is usually based on attempting to raise the cardiac output by giving large volumes of fluids intravenously, but there are dangers of overloading the circulation and upsetting the pH and electrolyte balance of the body since kidney function will probably be impaired.

Patient Monitoring. In addition to the clinical appraisal, it is usual to monitor the central venous pressure, the arterial blood pressure, the ECG as well as the plasma electrolytes, pH and blood gases.

In severe cases the shock is irreversible.

Sympathetic Activity and Exercise

In exercise the increased sympathetic activity causes a more powerful contraction of the ventricles (as well as an increase in heart rate), and, as a result, the ventricles empty more completely. By so doing they increase the stroke volume by using the **systolic reserve,** and, in effect, this is equivalent to a small blood transfusion. There is now more blood available to be pumped out for that beat, and since we are dealing with a circulation, that blood will ultimately return to the heart, and be pumped out again.

In addition the action of the muscle pump in the limbs, and vasodilatation of the muscle blood vessels by the metabolites, and the increase in breathing (**respiratory pump**) will increase the venous return of blood to the heart and further increase the stroke volume by Starling's law of the heart.

However, in many people much of the increase in cardiac output can be accounted for by the increase in heart rate.

The following summarizes some of the changes which occur in exercise.

1. An increase in heart rate and stroke volume leads to an increase in cardiac output.
2. The increased cardiac output coupled with a fall in peripheral resistance, due to the dilating action of metabolites on the muscle arterioles, gives a slightly raised systolic blood pressure but a normal or slightly lowered diastolic blood pressure. The pulse pressure (the difference between the two) is increased.
3. There is a redistribution of the blood flow so that an increased percentage of the cardiac output passes to the active muscles (and skin as body temperature rises) and a reduced percentage to the digestive tract.
4. More oxygen is extracted from each 100 ml. blood as it passes through the active muscles.
5. The rate and depth of respiration is increased. This is brought about by the action of higher centres, the increased carbon dioxide production, production of lactic acid, and the increased utilization of oxygen, on the respiratory centre [p. 55].
6. If the oxygen supply to the active muscles is insufficient for *aerobic metabolism*, energy is obtained by *anaerobic metabolism* and lactic acid is produced. Increased respiration continues after the exercise has finished. The additional oxygen which is then taken in, removes the lactic acid and repays the 'oxygen debt' [p. 68].

Measurement of Cardiac Output

Concentration is the amount of a substance in a given volume of fluid.

If, for example, 2500 units of a harmless dye are injected into a vein of a patient and the dye appears in the arterial blood for a total of 30 seconds with a mean concentration of 1 unit of dye per millilitre of blood, it follows that the dye must have met 2500 ml. of blood during its passage through the heart and lungs. (2500 units of dye in 2500 ml. of blood has a concentration of 1 unit of dye per ml. blood). It met the 2500 ml. blood in 30 seconds, so that the blood flow per minute through the heart and lungs will be double this, that is, the cardiac output is 5000 ml. per minute. This is the basis of the **dye-dilution technique** for the determination of cardiac output.

$$\text{Cardiac output} = \frac{\text{Amount of dye injected}}{\text{Concentration in arterial blood} \times \text{Time}}$$

As an alternative to the dye a radio-isotope may be employed and its concentration in the arterial blood measured using a radio-activity counter.

In the **Fick principle** method of determining cardiac output, the increase in the concentration of oxygen in the blood as it passes through the lungs is measured.

If a patient has an oxygen uptake of 250 ml. per minute, and each ml. of blood passing through the lungs takes up 0.05 ml. oxygen (or 5 ml. oxygen per 100 ml. blood), it follows that 250/0.05 = 5000 ml. blood must pass through the lungs per minute. Since we are dealing with a circulation, the lung blood flow is the same as the cardiac output.

$$\text{Cardiac output} = \frac{\text{oxygen uptake per minute}}{\text{arterial-venous oxygen difference}}$$

The oxygen in the blood leaving the lungs is obtained from an arterial blood sample. The determination of the oxygen in the blood arriving at the lungs involves a cardiac catheterization to obtain a sample of blood from the pulmonary artery. The difference in their oxygen concentrations (arterial-venous oxygen difference) gives the amount of oxygen taken up by each millilitre of blood passing through the lungs.

The oxygen uptake per minute is determined using a spirometer [pp. 48, 71].

Summary:

Cardiac output is determined by the heart rate and the stroke volume

$$\text{C.O.} = \text{H.R.} \times \text{S.V.}$$

1. Heart rate is modified by the vagus and sympathetic activity.

2. Stroke volume is the difference between the volume of the ventricles at the beginning of systole and the volume of the ventricles at the end of systole.

(a) The volume of blood in the ventricles at the beginning of systole depends on by how much the blood volume exceeds the capacity of the circulation, and this depends mainly on the venomotor tone. The greater the venomotor tone the greater the volume of blood in the heart at the beginning of systole.

(b) The volume of blood at the end of systole depends on the cardiac sympathetic activity.

3. In exercise part of the increase in cardiac output which occurs is due to a more complete emptying of the ventricles as a result of increased cardiac sympathetic activity.

6. BLOOD

An adult has about 5 litres of blood in his body. A new-born baby has only 300 ml. of blood. Blood makes up about one-twelfth of the body weight, and this fact may be used to estimate the blood volume of a baby, child, or an adult.

If blood is allowed to stand and prevented from clotting, it separates out into plasma and cells [FIG. 54].

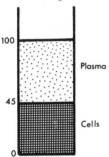

FIG. 54. If clotting is prevented, blood separates out into cells and plasma. Using an haematocrit tube, calibrated from 0 to 100, the cells occupy a volume of 45.

This process may be speeded up using a centrifuge [FIG. 55]. If this is carried out in a small tube calibrated from 0–100, the cells which settle at the bottom occupy a volume of 45 and the plasma occupies a volume of 55. Thus 45 per cent. of blood is made up of cells; the figure '45' is known as the **packed cell volume** or **haematocrit.** In anaemia, where there is a deficiency of red cells, the haematocrit is low.

BONE MARROW

The blood cells are made in the bone marrow. This is found in the hollow medullary cavity of bones. In the child all the bone marrow is engaged in cell formation, but in the adult, the active marrow, which is called **red bone marrow,** is restricted to the trunk and skull. Although it is termed red bone marrow it makes both

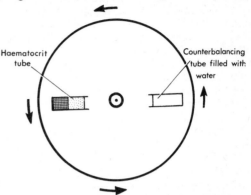

FIG. 55. The separation of blood into cells and plasma may be speeded up using a centrifuge. This works on the same principle as a spin-drier. The tube containing the blood must be counter-balanced. The centrifugal force produced by rotation at a high speed sends the denser cells to the periphery.

the red and the white blood cells. The marrow in the limb bones, the long bones of the body, is in the form of **yellow fatty marrow;** it is in reserve and is not actively engaged in making blood cells. It can, however, change to red bone marrow if there is a shortage of blood.

The bone marrow contains five different types of blood-forming cells. These are the **myeloblasts,** the **lymphoblasts,** the **monoblasts, erythroblasts,** and **megakaryocytes** [FIG. 56].

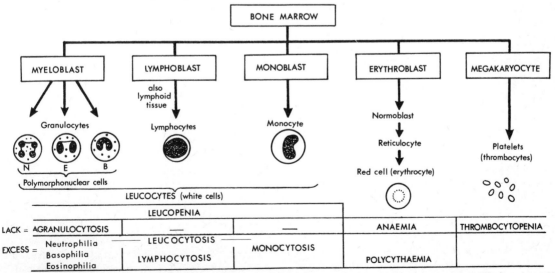

FIG. 56. Diagram showing the formation of blood cells by the bone marrow. N. = neutrophils, E. = eosinophils, B = basophils.

GRANULOCYTES

The myeloblasts give rise to the white blood cells that contain granules, these are known as **granulocytes.** In order to see these granules clearly, the cells are stained using dyes. The two dyes commonly employed are **eosin** (a red dye which is used for making red ink) and a **basic dye,** such as haematoxylin, which is blue in colour. When the granulocytes are stained using these dyes the granules present in the majority of granulocytes do not have a preference for either the eosin or the basic dye, and only stain lightly with a little of each dye. These cells are called **neutrophils.**

Certain of the granulocytes have granules which take up the eosin dye, and these granules become bright red in colour. These cells are termed **eosinophils.** A smaller number of granulocytes stain with the basic stain and these are termed **basophils.**

The nuclei of the granulocytes are divided into many parts, and these cells are therefore called **polymorpho-nuclear cells.** This lobulation of the nucleus, which is shown best in the neutrophils increases with the cell's age. The nucleus is divided, first into two lobes, then as the cell becomes older into 3, 4 and finally 5 lobes [FIG. 57].

FIG. 57. The nucleus of the neutrophils is divided into lobes. This lobulation increases from two to five as the cells age.

The lobulation in the case of the eosinophil is not quite so marked, and the nucleus is seldom divided into more than two lobes. In the case of the basophil, the nucleus shows a single constriction which divides the nucleus into two parts.

These granulocytes are **phagocytes,** that is, they have the power of ingesting foreign particles including bacteria. The granules contain digested enzymes which are able to dissolve the foreign particles.

LYMPHOCYTES

The lymphoblasts which are also found in the lymphoid tissue of the body, are the cells which give rise to the **lymphocytes.** The lymphocytes are slightly smaller cells than the granulocytes, and they have a very large spherical nucleus that occupies nearly all of the cell. These lymphocyte cells do not act as phagocytes. Instead they produce **antibodies,** which are able to react with foreign substances, known as **antigens,** and to destroy these antigens by an antigen-antibody reaction [p. 86].

MONOCYTES

The monoblasts are cells which give rise to the monocytes. These are larger white blood cells than the granulocytes or the lymphocytes. They usually have a kidney-shaped nucleus, and like the granulocytes, are phagocytic cells able to engulf bacteria and foreign particles.

White Cell Count

The granulocytes make up about 70 per cent. of the white cell count in the body (neutrophils 65 per cent., eosinophils 4 per cent., basophils 1 per cent.).

The lymphocytes make up about 25 per cent. of the white cells in the adult, and the monocytes the remaining 5 per cent. Children have much more lymphoid tissue than adults, and as a result, have many more lymphocytes than they have granulocytes.

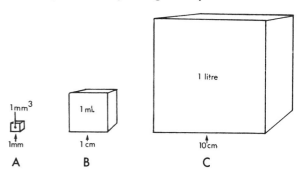

FIG. 58. A cube having sides 1 mm has a volume of 1 cubic millimetre (1 mm.^{3}) = 1 microlitre (μl).

B. A cube having sides 1 cm. has a volume of 1 ml. (= 1 c.c.). 1,000 cubes size A will fit into one cube size B. Thus 1,000 mm.^{3} = 1 ml.

C. A cube having sides 10 cm. has a volume of 1 litre. 1,000 cubes size B will fit into one cube size C. Thus 1 litre = 1,000 ml. = 1,000,000 mm^3.

A cubic millimetre of blood contains 8,000 white cells and 5,000,000 red cells.

The cubes illustrated are not to scale.

Altogether there are about 8,000 white blood cells per cubic millimetre of blood [FIG. 58(A)]. These white blood cells are also known as **leucocytes.**

Red Cell Count

The erythroblasts give rise to the red blood cells. There are many more red blood cells than white blood cells. The normal red cell count is five million cells per cubic millimetre [FIG. 58(A)].

Cell counts are usually expressed as the number of cells in one cubic millimetre of blood. The number of cells in 1 ml. blood will be 1,000 times greater. The number of cells in one litre of blood will be 1,000,000 times greater [FIG. 58] since a cubic millimetre is the same as a microlitre (μl).

Platelet Count

The megakaryocytes give rise to the platelets (thrombocytes). These platelets play an important part in the clotting of blood. There are about 250,000 platelets per cubic millimetre.

Cell Formation

Both red and white cells have to be continually replaced as the old cells wear out and die. The bone marrow has to be highly active the whole time in order to make up the losses which are continually occurring. Repeated cell divisions are needed to enable the bone marrow to maintain the white and red cell count.

These cell divisions are inhibited by ionizing radiation. This type of radiation is given out by X-rays, radio-isotopes and atomic reactors. An over-dose of ionizing radiation thus prevents the bone marrow from making the red and white cells and platelets. A similar disruption of cell formation, which is termed a **blood dyscrasia,** occurs in some patients who are given certain drugs which are benzene derivatives.

The failure to produce red blood cells is termed **aplastic anaemia.** A reduction in the circulating white blood cell count is termed **leucopenia.** A disappearance of the granulocytes from the circulation is termed **agranulocytosis.** A reduction in the formation of platelets (thrombocytes) by the megakaryocytes is called **thrombocytopenia.**

RED BLOOD CELLS

The red blood cells are biconcave discs, seven microns in diameter, and two microns thick [FIG. 59]. They contain the pigment **haemoglobin** which is bright red in

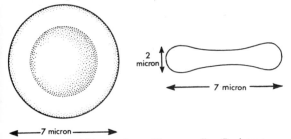

FIG. 59. Red blood cells are biconcave discs 7 microns in diameter and 2 microns thick. 1 micron (μ) = $\frac{1}{1,000}$ mm. = $\frac{1}{1,000,000}$ metre.

colour when combined with oxygen, and dark blue in colour when there is no oxygen present. Haemoglobin, as will be seen later, plays an important part in the carriage of both oxygen and carbon dioxide.

Every 100 ml. of blood has about 15 g. of haemoglobin in its red cells.

Dietary Requirements for Red Cell Formation

The red cells have a life of about 100 days, after which they are destroyed. This means that 1 per cent. of the red cells in the body have to be replaced each day. In order to do this we need certain items in the diet.

First we need protein to supply the amino acids, which are used as building bricks to form body protein, and in this case, the structure of the red blood cell.

In order to make the necessary haemoglobin, iron is needed in the diet. If the iron intake is not sufficient for the body's needs, an **iron deficiency anaemia** will develop.

The blood cells become very pale in colour due to the shortage of haemoglobin.

Much of the iron from the broken-down red cells is re-used in making the new cells. But an additional 5 mg. a day will be required by a man, and about 10 mg. a day by a woman. The reason why women need more iron than men is because they lose blood with each menstrual period, and this represents a complete loss of iron from the body. A patient who has had a haemorrhage may need an iron supplement in the diet to make up for the iron that was in the blood that has been lost.

In order to make a sufficient number of red cells, vitamin B_{12} and folic acid are required. Without vitamin B_{12}, folic acid metabolism is upset and too few red cells are made. Those that are made are irregular in size and in general are too big. This is termed a **megaloblastic** or **macrocytic anaemia.**

Such an anaemia is frequently associated with a leucopenia and thrombocytopenia since the vitamins play a part in the formation of white cells and platelets.

Vitamin B_{12} is a particularly difficult vitamin to absorb from the digestive tract into the blood. The bsorption is only adequate for the manufacture of red blood cells if the stomach produces a substance called the **intrinsic factor.** This is a protein–polysaccharide which aids the absorption of vitamin B_{12}. (Vitamin B_{12} was previously known as the **extrinsic factor.**)

After absorption the vitamin B_{12} is stored in the liver, and is sent to the bone marrow to aid the maturation of the red blood cells. These stores are adequate for several years.

If the stomach fails to produce the intrinsic factor, the vitamin B_{12} absorption is inadequate and a form of megaloblastic (macrocytic) anaemia known as **pernicious anaemia** develops. Since vitamin B_{12} is also stored in the liver of animals, a diet of liver was originally used to treat this condition. This has mainly been replaced by injections of vitamin B_{12}.

Another form of anaemia known as sickle-cell anaemia will be discussed later [p. 66].

Erythropoietin

Whenever there is a sustained shortage of oxygen in the body, the bone marrow responds by making more red blood cells. This is brought about by the release of a substance called **erythropoietin** by the kidneys in response to an oxygen lack.

Patients become anaemic following bilateral nephrectomies (removal of both kidneys).

RED CELL BREAKDOWN

The residue of the red cells at the end of their life is taken up by the reticulo-endothelial system of the body. This is the name given to a series of phagocytic cells which are found in the bone marrow, the spleen, the liver and in other parts of the body.

The haemoglobin is converted by these cells into the yellow pigment **bilirubin.** This change from the red haemoglobin to the yellow bilirubin can be seen in a bruise or in a black eye. The haemoglobin first loses its oxygen and becomes dark in colour. It then slowly changes first of all into the green substance biliverdin, and then finally to the yellow substance bilirubin.

Bilirubin passes via the blood to the liver where it is excreted into the bile, passes down the bile-duct into the duodenum, and then passes through the whole of the small and large intestines to appear in the faeces as the brown colouring matter **stercobilin** [FIG. 60].

This route of excretion is very important, and is also unusual, since most of the waste products in the body are excreted dissolved in water in the urine.

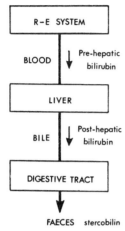

FIG. 60. The excretion of bilirubin. The bilirubin formed by the breakdown of red cells passes from the reticulo-endothial system (R-E system) to the liver via the blood. The liver extracts the bilirubin and passes it down the bile-duct as bile pigments to the duodenum. It is excreted in the faeces as brown stercobilin.

Jaundice

If this excretory pathway fails, and bilirubin accumulates in the blood, it will pass from the blood into the skin and produce a yellow coloration. This condition is known as **jaundice** or **icterus.** The yellow coloration will be seen particularly clearly in the eyes where the conjunctiva will become yellow in colour.

The excretion of bilirubin may be inadequate for three reasons. Firstly so much bilirubin may be produced that the liver is unable to transfer it sufficiently rapidly to the bile duct and jaundice develops. This may occur in the first few days of life when the baby's liver is immature. A baby is born with an excess of red cells (red cell count 7 million per cubic millimetre). The excess cells break down in the early days of life to reduce the red cell count to 5 million per cubic millimetre. The bilirubin from these broken-down red cells may be too much for the liver to deal with, and the baby becomes jaundiced. This is a **physiological jaundice.** It should not be confused with a more severe form of jaundice which is the result of a Rhesus factor mismatch.

A similar form of jaundice due to the too great a production of bilirubin can occur in certain forms of

malaria, and in a congenital condition called **acholuric jaundice.** In acholuric jaundice the patient has episodes when the red blood cells disintegrate and release a large amount of haemoglobin which is then changed into bilirubin.

A second type of jaundice occurs when the liver is diseased, for example by infective hepatitis. This is a virus infection which causes jaundice since it prevents the liver cells from transferring the bilirubin from the blood to the bile duct.

The third form of jaundice is that due to an obstruction to the bile-duct. Biliary obstruction may arise either from an impacted gall-stone [FIG. 61], or by pressure on the bile-duct from a carcinoma of the pancreas. To treat this type of jaundice it is necessary to remove the obstruction to the bile-duct.

If the bilirubin is no longer reaching the intestines, the faeces become pale in colour (clay-coloured stools). The bilirubin in these conditions is excreted in the urine and as a result the urine becomes very much darker in colour.

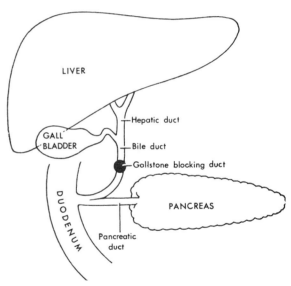

FIG. 61. The excretion of bilirubin into the duodenum may be blocked by a gall-stone in the bile duct. This is one of the causes of jaundice.

INVESTIGATION OF ANAEMIA

An anaemia due to a deficiency of iron (iron deficiency anaemia) is associated with a deficiency of haemoglobin in the red cells. An anaemia due to a deficiency of vitamin B_{12} and folic acid (macrocytic anaemia) is associated with a deficiency of red cells, but the cells that are present will be larger than normal and will contain an adequate amount of haemoglobin.

In order to differentiate between these two types of anaemia, a red cell count, a haemoglobin estimation, and a haematocrit determination are made.

Mean Corpuscular Volume

The average volume of a red cell is found by dividing the volume of red cells in 100 ml. blood by the number of cells in 100 ml. blood. This is termed the mean corpuscular volume (M.C.V.).

$$\text{M.C.V.} = \frac{\text{Haematocrit}}{\text{No. of red cells}}$$

Normally, $\text{M.C.V.} = \dfrac{45}{5,000,000 \times 10^5} = \dfrac{90}{10^{12}}$ ml.

$$= 90 \text{ cubic microns}$$

where 1 micron $= \dfrac{1}{1,000}$ mm. $= \dfrac{1}{10,000}$ cm.

A higher than normal red cell volume indicates that the red cells are larger than normal, i.e. macrocytosis.

Mean Corpuscular Haemoglobin

The average amount of haemoglobin in a red cell is then determined by dividing the haemoglobin in 100 ml. blood by the number of cells in 100 ml. blood. This is termed the mean corpuscular haemoglobin (M.C.H.).

$$\text{M.C.H.} = \frac{\text{Haemoglobin}}{\text{No. of red cells}}$$

Normally,

$$\text{M.C.H.} = \frac{15 \text{ g./100 ml.}}{5,000,000 \times 10^5/100 \text{ ml.}} = \frac{30}{10^{12}} \text{ g.}$$

$$= 30 \text{ micromicrograms} = 30 \text{ picograms.}$$

(There are 5,000,000 red cells per cubic millimetre but 100,000 times this number ($\times 10^5$) per 100 ml.)

Mean Corpuscular Haemoglobin Concentration

The mean corpuscular haemoglobin suffers from the disadvantage that it does not take into account the cell size. A better indication of the haemoglobin content of a red cell is found by dividing the haemoglobin content by the haematocrit. This gives the quantity of haemoglobin in a given volume of red cells, i.e. it is a concentration. It is termed mean corpuscular haemoglobin concentration (M.C.H.C.) and is expressed as a percentage.

$$\text{M.C.H.C.} = \frac{\text{haemoglobin}}{\text{haematocrit}} \times 100\%$$

Normally

$$\text{M.C.H.C.} = \frac{15}{45} \times 100\%$$

$$= \tfrac{1}{3} \times 100\% = 33\%.$$

Multiplying the grams of haemoglobin per 100 ml. blood by 100 and dividing by the haematocrit gives the *grams of haemoglobin* present in *100 ml. packed red cells*. A mean corpuscular haemoglobin concentration of 33 per cent. means that there are 33 grams of haemoglobin in every 100 ml. of packed red cells.

A low M.C.H.C. is found in an iron deficiency anaemia.

BLOOD TRANSFUSIONS

If a patient becomes severely anaemic it may be no longer possible to treat this by simply improving the diet and to add iron, B_{12} or folic acid supplement. It may be necessary to give a blood transfusion.

The blood for a blood transfusion is taken under sterile conditions. Acid citrate is added to prevent the blood from clotting, and glucose (dextrose) added to feed the red cells (ACD = acid citrate dextrose blood, CPD = citrate, phosphate *as buffer*, dextrose blood). If the blood is to be used in a heart-lung machine, then *heparin* is used as the anticoagulant to prevent the blood clotting.

The white cells do not survive for more than a few hours when the blood is taken for a blood transfusion and a transfusion of stored blood can therefore not be used to increase the number of white cells in the circulation.

It is very important that none of the blood transfusion apparatus should contain any soap or detergent because these substances dissolve the outer fatty membrane of the red blood cell, and will thus destroy the cells.

The blood is stored at 4°C. and care must be taken to ensure that the temperature does not fall below freezing point because this would destroy the red blood cells, due to the formation of ice crystals.

Experiments are being carried out to find methods by which blood can be stored indefinitely. The addition of glycerol or transfusible colloids such as hydroxyethyl starch and polyvinyl-pyrrolidone enable the blood to be deep-frozen without cell rupture by ice crystals. The glycerol has to be extracted from the blood after thawing.

Blood Groups

The red cells have on their outer coat a sugary substance called an **agglutinogen**. Nearly half the population (42 per cent.) have agglutinogen **A** on every single red cell in their body and these people are said to be **Group A**. A smaller percentage (9 per cent.) have agglutinogen **B** on their red cells and they are said to be **Group B**. About 3 per cent. of the population, have both substances **A** and **B** on their cells and they are said to be **Group AB**. The remainder of the population (46 per cent.) have neither **A** nor **B** on their red cells, and they are said to be **Group O** [FIG. 62].

During the first year of life the people who are **Group A** develop in their plasma a substance which destroys any cell that has substance **B** on it. Exactly how this is brought about is not fully understood, but is possible that some of the bacteria which enter the digestive tract have a similar sugar-like substance on their outer coat to agglutinogen **B** and the development of the **Anti-B** is the body's response (or antibody formed) to the antigen **B**.

Group A develop **Anti-B** in their plasma, **Group B** develop **Anti-A** in their plasma, and **Group O** develop **Anti-A** and **Anti-B** in their plasma. It is only the **Group**

AB who do not develop any **agglutinins,** as they are called, in their plasma.

When **Group A** red blood cells come in contact with plasma that contains **Anti-A,** they become sticky and clump together. This is termed **agglutination.**

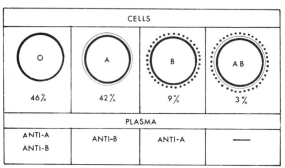

FIG. 62. Blood groups.

Agglutination is particularly likely to occur in the small blood vessels of the kidney, and such agglutination will stop the kidney from functioning which gives rise to failure to produce urine which is known as **anuria.** It is therefore very important that no person who is **Group A** should ever be given **Group B** blood, nor should any person who is **Group B** ever be given **Group A** blood.

A person who is **Group O** and whose plasma contains **Anti-A** and **Anti-B** should not be given either **A** blood, or **B** blood or **AB** blood but must only have their own blood group.

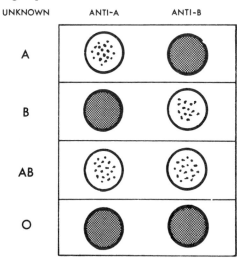

FIG. 63. Determination of an unknown blood group. A very small quantity of blood is added to ANTI-A and ANTI-B serum on a slide. The slide is rocked and inspected for agglutination (as top left). The agglutination pattern gives the blood group.

Blood transfusions are carried out on the basis that each patient should have his own blood group, and only if this group is not available, in special cases, would **Group O** blood be used.

It is possible to detect the agglutination of blood outside the body [FIG. 63] and a crossed-match should be carried out between the blood that is about to be given to a patient, and the patient's own plasma to ensure that the cells about to be given will not be agglutinated in the circulation.

Rhesus Group

Up to 1940, the **ABO** blood group system was the only system of importance that was known. Then Landsteiner (who had discovered the **ABO** grouping in 1900) discovered that another substance **D** was of importance and should be taken into consideration. Patients who have agglutinogen **D** on their red cells are said to be **Rhesus positive,** and about 85 per cent. of the population are in that category. The remaining 15 per cent have no **D** on their red cells, and are said to be **Rhesus negative.**

Neither the Rhesus positive nor the Rhesus negative people have any **Anti-D** naturally occurring in their plasma, but **Anti-D** will develop in a Rhesus negative person only, if that person is given a Rhesus positive transfusion. Having once been given such a transfusion, the **Anti-D** which is formed will persist throughout life, and on all subsequent occasions that person will not be able to have Rhesus positive (**D**) blood but will have to have his own correct Rhesus group (Rhesus negative blood). This only applies to the Rhesus negative patients. The Rhesus positive patients do not have **Anti-D** in their plasma, they cannot make **Anti-D** and they can have a transfusion of either Rhesus negative or Rhesus positive blood.

As far as males are concerned, if such a patient is Rhesus negative, and is given Rhesus positive blood by mistake, the only untoward thing that will happen will be that **Anti-D** will develop, and from then on transfusions will have to be restricted to Rhesus negative blood.

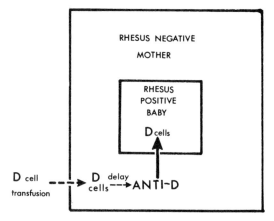

FIG. 64. If D cells (Rhesus positive) enter a Rhesus negative woman's circulation, after a delay, ANTI-D will develop. This ANTI-D will cross the placental barrier at a subsequent pregnancy and destroy the baby's cells if these are Rhesus positive. The ANTI-D will have no effect if the baby is Rhesus negative.

In the case of females, no woman of child-bearing age or younger should ever be allowed to develop **Anti-D** because this may prevent her from having a living child. The reason for this is that the Rhesus factor is inherited from one's parents, and many Rhesus negative women have Rhesus positive children because their husbands are Rhesus positive. This does not in itself matter because the baby's circulation is independent of the mother's. It has its own heart, its own blood, its own blood vessels, and although its blood comes close to the mother's blood in the placenta, the cells of the baby, under normal conditions, do not enter the mother's circulation. So it is a comparatively common occurrence that a Rhesus negative mother has inside her uterus a Rhesus positive baby.

If the mother has at any time in her life developed **Anti-D**, this **Anti-D** is able to pass across the placental barrier and enter the baby's circulation, where it will destroy the baby's cells [FIG. 64]. When this occurs, the only method known at the present moment to save the baby's life is to carry out an exchange transfusion, and to change the baby's blood temporarily from Rhesus positive to Rhesus negative. Rhesus negative blood is unaffected by **Anti-D** and although the baby will revert back to its own blood group (Rhesus positive) over the course of the next few weeks, its life will have been saved. These exchange transfusions have been carried out *in utero;* more commonly, they are carried out as soon as the baby is born. If an exchange transfusion *in utero* is not possible, the baby may die and the mother will have a miscarriage.

Some women develop **Anti-D** after one or more pregnancies with Rhesus positive babies, although they have not had any blood transfusion of Rhesus positive blood. What has probably happened in such a case is that placental damage has let some of the baby's blood, which is Rhesus positive, through into the maternal circulation where it has acted as a blood transfusion for the mother with the wrong type of blood. **Anti-D** then develops in the mother. This is particularly likely to happen during child-birth when the placenta separates from the mother's uterus leaving a large raw area. The placenta contains the baby's Rhesus positive blood, and some of the blood may get into the mother's circulation at this time.

It now appears from recent clinical trials that this iso-immunization of Rhesus negative women can be largely prevented by the injection of anti-D gamma globulin shortly after delivery.

PLATELETS

The platelets, or thrombocytes, have two main functions in the body. The platelets are able to clump together and by so doing to block small holes in blood vessels. This is termed a platelet plug, and it plays an important part in the arrest of a small haemorrhage such as is caused by a pin prick in *bleeding time* test.

When a larger haemorrhage occurs, or when there is an extensive injury the blood clots.

BLOOD CLOTTING

Blood clotting is a change which occurs in the plasma. The red and white cells play no part in the formation of a blood clot, although they may be trapped in the clot when it forms. The first stage in the clotting of blood occurs when the platelets break down, or when tissues are damaged. Either of these events gives rise to the formation of a substance called **thromboplastin** (thrombokinase). Thromboplastin, when it is formed, converts a substance which is present in the plasma known as **prothrombin,** into **thrombin.** This change will only take place if there are calcium ions in the blood. Thrombin, when it is formed, acts on another substance in the plasma known as **fibrinogen** and changes it into **fibrin.** It is the formation of fibrin which gives rise to the clot [FIG. 65].

The clot when it is first formed is soft and jelly-like, but after a short time it contracts down, and exudes a straw-coloured fluid which is called **serum.** If blood is prevented from clotting, it separates out into **cells and plasma**; if it is allowed to clot it separates out into a **clot plus serum.** Serum and plasma are very similar, the only difference being that serum has lost the clotting factors prothrombin and fibrinogen which are present in the plasma.

Patients with haemophilia, or with Christmas disease, are unable to make thromboplastin even though their platelets break down. This is due to an inherited defect in the materials in the plasma that are needed for the formation of thromboplastin. These missing factors are known as anti-haemophilic globulin (AHG or Factor VIII) and Christmas factor (Factor IX).

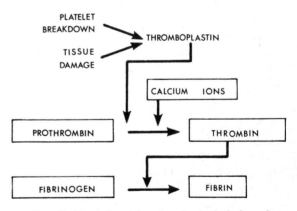

FIG. 65. Blood clots when thromboplastin is formed by the breakdown of platelets or damage to tissue.

Blood clotting may be prevented in several ways. The method used when blood is taken for a transfusion is to remove the calcium ions by adding sodium acid citrate to the blood. This removes the calcium ions as sodium calcium citrate. The blood will then not clot even though thromboplastin is formed since this thromboplastin will be unable to change the prothrombin into thrombin.

An alternative way to prevent the blood from clotting is to use heparin which is a naturally occurring substance produced by the **mast cells.** These cells are found lining certain blood vessels in the body. Heparin prevents the prothrombin changing into thrombin, and prevents the fibrinogen changing into fibrin.

Should clotting occur in the blood vessels of the body, this clot or **thrombus** gives rise to a **thrombosis.** Such a clot may have very serious consequences if it occurs in the blood vessels to the heart (coronary thrombosis) or in the blood vessels to the brain (cerebral thrombosis). If a thrombus occurs in a vein such as one of the veins of the legs, it may become detached and form an **embolus.** An embolus arising in the veins of the legs would finally become lodged in the blood vessels to the lungs, and give rise to a **pulmonary embolus.**

Oral anticoagulants, such as the derivatives of *dicoumarol,* lower the blood prothrombin level by inhibiting its synthesis in the liver. This increases the blood clotting time and reduces the likelihood of intra-vascular thrombus formation.

THE PLASMA

Plasma is the straw-coloured fluid in which the cells are suspended. It consists of a watery solution of the plasma proteins and the plasma electrolytes, plus all the substances which are being transported by the blood.

The plasma proteins are subdivided into the plasma **albumin** and the plasma **globulin,** the globulin being larger molecules than the albumin. The globulin may be further subdivided according to the size of the molecules, and the first three letters of the Greek alphabet are used to denote this, the globulins being termed alpha, beta and gamma globulins, respectively.

Every 100 ml. plasma has about 4·5 g. albumin, and about 2·7 g. of globulin. These plasma proteins are mainly manufactured in the liver, although some of the globulins are manufactured by the lymphocytes and lymphoid tissue.

In liver disease there is a reduction in the formation of albumin, and the globulin may in fact exceed the amount of albumin in the plasma.

The principal electrolytes in the plasma are sodium chloride and sodium bicarbonate. The sodium bicarbonate is very important in order to keep the blood slightly alkaline. The sodium chloride is present in plasma to the extent of 0·9 g. per 100 ml. A solution containing this amount of sodium chloride in water is termed **normal, isotonic,** or **physiological** saline, and it has the same electrolyte strength as blood. Such a solution, if sterile, could be run into the veins of a patient, whereas pure distilled water would destroy the red blood cells by the process of **haemolysis.** This is because the red blood cell contains a strong salt solution of potassium salts, and if it is placed in distilled water, it sucks in water by the process of **osmosis** [p. 41].

Function of Plasma Proteins

The plasma proteins have a number of functions.

1. They act as a protein reserve to the body, and can be used to supply body protein in states of starvation.

2. They exert an osmotic pressure of 25 mm. Hg, which, as will be seen in the next chapter, plays an important part in the formation and reabsorption of tissue fluid.

3. They are important in transporting certain hormones and other substances in the blood, particularly substances which would not otherwise be soluble. An example of such transport is the movement of bilirubin from the reticulo-endothelial system round to the liver. It is transported combined with plasma protein.

4. The plasma proteins increase the stickiness (viscosity) of the blood and this enables a blood pressure to be developed.

5. They also have the ability to neutralize both acids and alkalis, that is, they act as a buffer.

6. The circulating antibodies are found principally in the gamma globulin fraction of the plasma proteins [p. 86].

THE SPLEEN

The spleen in man is not essential to life. It may be removed (splenectomy) without harmful effects.

In animals, the spleen acts as a store of red blood cells, and during muscular activity or following a haemorrhage, these cells are discharged into the blood. In man the spleen is too small for any expelled red cells to have a significant effect on the total number of red cells in the body.

The spleen is the site of destruction of red blood cells and a site of formation of lymphocytes. It is part of the reticulo-endothelial system, and removes foreign particles including broken-down red cells from the circulation. In malaria the spleen enlarges enormously and may fill a major part of the abdominal cavity.

The spleen is a site of destruction of platelets (thrombocytes). Over-activity of the spleen (hypersplenism) may lead to a platelet deficiency (thrombocytopenia) which is associated with spontaneous haemorrhages from capillaries (purpura). Such a condition is improved by splenectomy.

7. TISSUE FLUID AND LYMPH

THE FORMATION OF TISSUE FLUID

The blood flowing through blood vessels does not come in contact with any cells of the body. It stays in the blood vessels and the nearest blood vessel to a body cell is usually a capillary. The space between the capillary and the cell is filled with tissue fluid [FIG. 66]. The food

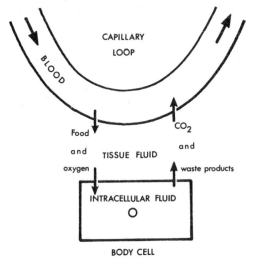

FIG. 66. Food and oxygen diffuse from the blood through the tissue fluid to the body cells. Carbon dioxide and waste products diffuse in the reverse direction from the body cell to the tissue fluid and then into the blood.

and oxygen have to diffuse through the tissue fluid from the capillary to the cell. Waste products diffuse backwards from the cell through the tissue spaces back to the capillary.

Diffusion

If a strong solution is separated by a membrane from a weaker solution [FIG. 67(A)], the substance dissolved (solute) will pass from the strong solution to the

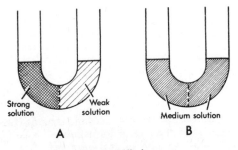

FIG. 67. Diffusion.

weaker until both solutions have the same strength [FIG. 67(B)]. This movement of a solute from a strong solution to a weaker solution is termed **diffusion.**

Osmosis

If a strong solution is separated by a membrane from a weaker solution but the membrane will not allow the solute molecules to pass through [FIG. 68(A)], water

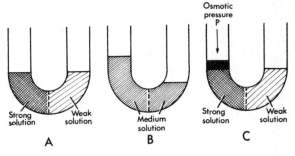

FIG. 68. Osmosis.

moves in the opposite direction until both solutions have the same strength [FIG. 68(B)]. This movement of water from the weak solution to the stronger solution is termed **osmosis.**

This movement of water can be prevented by applying a hydrostatic pressure to the strong solution [FIG. 68(C)]. The pressure required to prevent the strong solution 'sucking' in the water from the weaker solution is the osmotic pressure. It is measured in millimetres of mercury.

The capillaries behave as such a membrane with respect to the proteins in the plasma. The capillaries are permeable to water, but impermeable to the plasma proteins. This fact is important in the formation of tissue fluid.

EXTRACELLULAR FLUID

Water makes up about 70 per cent. of the total body weight, and a 70 kg. man has about 45 litres of water in his body. Thirty litres of this water is found inside the cells, and is termed **intracellular fluid** (*intra* L. = inside). The salts in this intracellular fluid are mainly potassium and magnesium salts.

The fluid that is outside the cells, namely 15 litres, is termed the **extracellular fluid** (*extra* L. = outside), and the electrolytes in this fluid are mainly sodium salts. There is very little sodium in the cells of the body because the cells have a **sodium pump,** and pump out any sodium that diffuses in. The extracellular fluid is divided into the extracellular fluid of the blood, which is known as **plasma,** and the remaining extracellular fluid of the body which is termed interstitial fluid (*inter* L. = between). The interstitial fluid is also referred to as **tissue fluid**, and this latter term will be employed.

TISSUE FLUID FORMATION AND REABSORPTION

It has been seen that by the time the capillary is reached, the blood pressure has fallen to 32 mm. Hg [FIG. 49, p. 27]. It falls further to 12 mm. Hg as the blood passes along the capillary.

The capillary membrane is permeable to water, and the blood pressure in the capillary will tend to drive water out through the capillary wall into the tissue spaces, thus causing the formation of tissue fluid. Tissue fluid contains virtually no plasma protein, and the presence of plasma proteins in the blood produces an osmotic pressure which sucks water in from the tissue spaces. The magnitude of this suction force is 25 mm. Hg. (This is the same force as is used when drinking through a straw). This osmotic suction force is in opposition to the capillary blood pressure. The pressure is forcing water out; the osmotic suction force is sucking water in.

At the arteriole end of the capillary, the pressure of 32 mm. Hg forcing fluid out, is greater than the osmotic pressure force of 25 mm. Hg sucking fluid in [FIG. 69].

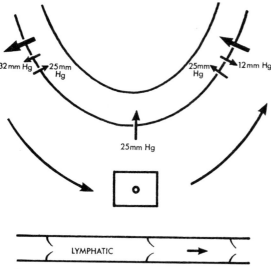

FIG. 69. Formation and reabsorption of tissue fluid. The presence of plasma proteins in the blood capillary and their absence in the tissue fluid provides a suction force of 25 mm. Hg which exceeds the blood pressure at the venous end of the capillary and returns fluid to the circulation. Any plasma proteins which leak from the capillary into the tissue spaces are taken up by the lymphatics.

As a result more fluid is formed than reabsorbed, and the net result is that tissue fluid is formed at this part of the capillary. At the venous end of the capillary, the osmotic suction force is greater than the capillary blood pressure, and, at this point, the net result will be the return of tissue fluid to the capillary.

Under normal conditions there is equilibrium between the formation of tissue fluid and the reabsorption of tissue fluid. The reabsorption of tissue fluid, however, depends upon the tissue spaces remaining free of plasma protein, since the osmotic suction force would disappear if proteins were allowed to accumulate in the tissue fluid. Although the capillary membrane can be considered to be impermeable to protein, a small amount of protein does leak out from the capillaries over the course of time. This protein is removed by means of another system of tubes which run through the tissue spaces, known as **lymphatics**. These vessels pick up the plasma protein, and transfer it to their lumen. The

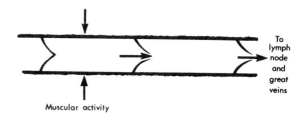

LYMPHATICS

FIG. 70. Lymphatics. The lymphatics contain valves and the alternate contraction and relaxation of the surrounding muscles forces the lymph along the lymphatic vessels.

lymphatics have valves, and like the veins, muscular contraction causes a flow of lymph along these vessels [FIG. 70]. Ultimately it is discharged into the great veins in the region of the neck.

Tissue Fluid and Restoration of Plasma Volume after Haemorrhage

If a haemorrhage is of sufficient magnitude to cause a fall in arterial blood pressure, the blood pressure in the capillaries will also be reduced. There will now be a lower blood pressure at both the arterial and venous ends of the capillary, and as a result the reabsorption of tissue fluid will now exceed its formation. The net result is that fluid will be transferred from the tissue spaces to the blood, thus tending to restore the plasma volume.

Even if the arterial blood pressure is maintained close to its previous level following the haemorrhage by the baroreceptors, fluid will still be restored to the blood from the tissue space. There will be a profound vasoconstriction of the arterioles and this will increase the pressure drop across these vessels. As a result the capillary blood pressure will be reduced and tissue fluid reabsorption will exceed production.

OEDEMA

An excessive amount of tissue fluid is termed **oedema**. It occurs when the rate of formation of tissue fluid is greater than the rate of reabsorption. Referring to FIGURE 69, it will be seen that the reabsorption depends on the osmotic pressure of the plasma proteins. Normally there are 7 g. of plasma proteins in every 100 ml. plasma, and this amount of protein is needed to give the 25 mm. Hg osmotic suction force [FIG. 71].

Plasma Protein Deficiency. A shortage of plasma proteins can occur in a variety of clinical conditions. A shortage may be due to an insufficient intake of food protein, a deficiency of manufacture of the plasma proteins or an excessive loss. Thus oedema, due to a deficiency of plasma proteins, may occur in conditions of starvation and malnutrition (failure of intake) [p. 72], or in liver disease, since the plasma proteins are mainly made in the liver (failure of manufacture), or in forms of nephritis (kidney disease), particularly the nephrotic syndrome, in which there is a large loss of protein in the urine (albuminuria). All these three clinical conditions are associated with oedema.

It will be noted [FIG. 71] that the reabsorption of

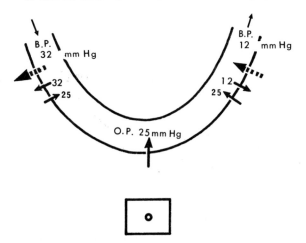

FIG. 71. Reabsorption of tissue fluid will only take place at the venous end of the capillary if the venous pressure is less than the osmotic pressure of the plasma proteins (25 mm. Hg).

tissue fluid can only occur if the blood pressure at the venous end of the capillary is lower than the osmotic pressure of the plasma proteins. Normally this is so because the capillary blood pressure is 12 mm. Hg at this point whilst the osmotic suction force is 25 mm. Hg. This blood pressure of 12 mm. Hg is the pressure available to pump the blood back to the right side of the heart via the veins. This pressure is adequate provided that there is no venous obstruction, and that the pressure at the right atrium is atmospheric pressure, namely 0 mm. Hg.

High Venous Pressure. If a venous obstruction has occurred, or if the patient is in congestive heart failure with a high venous blood pressure in the great veins and right atrium, then the pressure at the venous end of the capillary will be higher than 12 mm. Hg. If it is in

excess of 25 mm. Hg there will be no reabsorption of tissue fluid and oedema will develop.

A **venous thrombosis,** a very tight bandage, or a plaster cast which is obstructing the veins, will give rise to oedema distal to the point of the obstruction. Pregnant women may have swollen legs as a result of pressure by the baby *in utero* on the veins returning blood from the legs. Patients with **congestive heart failure** will have gross oedema due to the great increase in capillary blood pressure which results from the very high pressure in the right side of the heart.

Increased Capillary Permeability. Oedema also occurs if the capillaries become permeable to the plasma proteins, and the plasma proteins leak out into the tissue spaces. In congestive heart failure in addition to the high venous pressure there is generalized stagnant anoxia (oxygen lack), which damages the capillaries and as a result they become permeable to protein. Local capillary damage occurs following the release of histamine in the triple response [p. 21], and the wheal associated with this response is a localized form of oedema.

Oedema due to capillary damage is seen in injured areas such as sprained ankles, bruises, etc. where the localized oedema produces a swelling. Excessive swelling may be prevented by applying a tight bandage which prevents more tissue fluid being formed. Insect bites such as mosquito, gnats and fleas, produce capillary damage, and the raised area surrounding the bite is due to a localized oedema.

Blocked Lymphatics. Oedema also occurs if the lymphatics become blocked, are surgically removed or are congenitally absent (lymphoedema of the legs). The filaria worm is a parasite which lives in the lymphatics. Filaria infestation is often associated with very gross oedema of a limb or a part of the body. Oedema of the arm may develop following a radical mastectomy operation, which included the removal of the lymphatics.

Oedema Due to Sodium Retention. It will be seen later that generalized oedema may also occur in patients with sodium retention due to an excess of adrenal cortex hormones [p. 104]. Sodium chloride is retained in the body as an aqueous solution of 0·9 g. sodium chloride/ 100 ml. water which has the same osmotic pressure as the blood cells.

Sites of Oedema

Because of gravity, oedema fluid always tends to form in the dependent parts of the body. In the standing position, it accumulates in the ankles and legs. When a patient is in bed, the lowest part of the body is the lower sacral region, and oedema is often found here. A patient who is lying flat is more likely to show oedema in the face, than a patient who is sitting or standing.

Pulmonary Oedema

In the lungs the pulmonary arterial blood pressure is usually so low (25/8 mm. Hg) that all the pulmonary blood vessels including the pulmonary capillaries have a pressure of less than the osmotic suction force of the plasma proteins. As a result tissue fluid is not normally formed in the lungs. However, in any condition where the pressure in the left atrium and pulmonary veins is elevated this increase in pressure will be transmitted back to the pulmonary capillaries and pulmonary artery. The pressure in the pulmonary capillaries may rise to such an extent that tissue fluid will be produced causing **pulmonary oedema.** Since there are very limited tissue spaces in the lungs, this tissue fluid will enter the alveolar air sacs, and will cause dyspnoea (difficulty in breathing). The patient will become breathless, and the condition is referred to as **cardiac asthma.** The patient finds it easier to breathe in the sitting-up position. The blood pressure in the upper parts of the lungs in this posture is lower than that in the rest of the lungs and may be low enough for normal breathing to continue whilst the oedema fluid will collect at the bases of lungs [see gravity and blood pressure, p. 25].

HEART FAILURE AND OEDEMA

When the heart is unable to transfer blood from the venous side to the arterial side the condition is termed **heart failure.**

Left-sided Failure

Failure of the left ventricle may arise as a result of hypertension, coronary artery disease, or disease of the aortic valve.

In **left ventricular failure** the blood is not transferred adequately from the pulmonary veins through the left side of the heart to the aorta. As a result there is a build up of pressure in the pulmonary veins, and this pressure will be transmitted backwards through the lungs to the pulmonary artery.

The pulmonary capillary pressure will be higher than normal, and should it exceed 25 mm. Hg, tissue fluid will be formed in the lungs producing pulmonary oedema. It interferes with the exchange of gases, and makes the lungs stiffer [p. 60].

Left ventricular failure leads to dyspnoea which is much worse when lying flat. In this position there is more blood in the lungs. The patient can only breathe comfortably when sitting up. This is termed **orthopnoea.**

In the sitting-up posture the tissue fluid will collect at the base of the lungs. Breathing and exchange of gases will still be possible at the apices and in the upper parts of the lungs.

Right-sided Failure

If the right ventricle fails to transfer blood adequately from the great veins through to the pulmonary artery, the condition is termed **right ventricular failure.**

This may be the result of stenosis of the mitral valve, pulmonary hypertension, or following an embolus which has lodged in the pulmonary artery (pulmonary embolism).

Failure of the right side of the heart leads to an increase in pressure in the great veins, and as a result the neck veins become distended with blood.

Normally all veins above the sternal angle (manubrio-sternal junction) are collapsed because the pressure inside the veins is less than atmospheric. Thus in the sitting or standing posture the neck veins are not normally visible. In right ventricular failure, however, the pressure is raised to such an extent, that the neck veins are distended when sitting or standing.

The height of the column of blood in the veins above the sternal angle gives a measure of the increased pressure which exists on the venous side leading to the right heart.

The increase in venous pressure which is transmitted backwards through the veins to the capillaries of the systemic circulation will lead to generalized oedema.

Sodium Chloride and Water Retention in Heart Failure

In both types of heart failure the ventricles fail to maintain an adequate output, and as a result there is **stagnant anoxia** [p. 56].

The effect of the poor blood flow to the kidneys leads to the release of **renin,** and the formation of **angiotensin** which stimulates the adrenal cortex to release **aldosterone** [p. 104].

As a result there is retention of sodium chloride and water.

The urinary volume decreases. The urine has a higher specific gravity and may contain albumin.

The increase in blood volume brought about by the retention of sodium chloride and water, causes a further increase in venous pressure which makes the heart failure worse. It is necessary to break the vicious circle by reducing the blood volume. This may be achieved by venesection or by means of diuretics. Digitalis may also be used to increase the activity of the heart.

Thus heart failure is associated with a low cardiac output, stagnant anoxia and an increase in blood volume with sodium retention. To limit this sodium retention a salt-restricted diet is usually employed.

8. RESPIRATION

Respiration or breathing has three main functions:

1. To take in oxygen,
2. To give off carbon dioxide,
3. To regulate the pH of the blood.

The Need for Oxygen

Heat and energy are produced in the body by the oxidation of the carbon and hydrogen in the food. The oxygen required comes from the inspired air.

The conversion of food into heat and energy in the cells of the body takes place in a number of stages. The whole process, termed metabolism, is complex, but the over-all changes are that the carbon atoms from the food are combined with oxygen atoms from the air with the release of heat and energy and the formation of carbon dioxide (CO_2) as a waste product. In addition the hydrogen atoms from the food are combined with oxygen atoms from the air with the release of further heat and energy and the formation of the water (H_2O) as a waste product [FIG. 72].

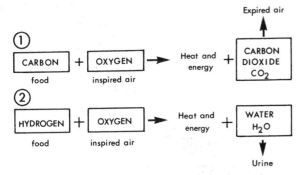

FIG. 72. Basic plan for the production of heat and energy.

The carbon dioxide is excreted in the expired air. The water formed (metabolic water) augments the water intake of the body [p. 78]. The surplus is excreted by the kidneys in the urine.

At rest, 250 ml. of oxygen per minute are absorbed from the inspired air to satisfy the metabolic requirements of the body. This figure is greatly increased when exercise is taken and in very severe exercise the oxygen requirement may be as high as 5,000 ml. of oxygen per minute.

Since some of the oxygen is used in the oxidation of hydrogen, the volume of carbon dioxide given off each minute is usually slightly less than the volume of oxygen taken in. Thus at rest only 200 ml. of carbon dioxide are given off for every 250 ml. of oxygen taken in.

The ratio of 'carbon dioxide given off' to 'oxygen taken in' is known as the respiratory quotient (R.Q.). It is normally $0.8 = \frac{200}{250}$.

The respiratory quotient gives an indication of the food being metabolized. If carbohydrate only is being used, the R.Q. equals 1 and the amount of oxygen taken in, and the amount of carbon dioxide given off are the same.

If fat is being metabolized, the R.Q. is 0.7 and every 250 ml. of oxygen taken in will give rise to only $\frac{7}{10}$ of 250 ml. of CO_2 (= 175 ml.).

On a mixed diet when fats, carbohydrates and proteins are being metabolized, as has been seen, the **R.Q.** = 0.8.

BREATHING

The lungs completely fill the thoracic cavity and breathing is brought about by increasing and decreasing the size of this cavity.

The air in the lungs is termed **alveolar air.** It differs from room air in that it contains much less oxygen and much more carbon dioxide. It is the alveolar air that comes in contact with the blood flowing through the pulmonary capillaries. (The blood in these vessels never comes in contact with room air.)

When analysed, alveolar air is found to contain:

14% oxygen
5.5–6% carbon dioxide
80% nitrogen

Room air contains:

21% oxygen
79% nitrogen
0% carbon dioxide.

(There is a very small amount of carbon dioxide in room air which is of great importance to plants, but this is only 0.03 per cent., and as far as man and animals are concerned this is equivalent to zero.)

At the end of a quiet expiration, the lungs still contain about 3 litres of alveolar air but only 420 ml. of this is oxygen (14 per cent. of 3,000 ml.). Since the body requires 250 ml. of oxygen every minute for metabolism, the reserve of oxygen in the lungs is less than a two minute supply. It is essential, therefore, to continually refresh the air in the lungs by the process of breathing. This process must be maintained without a break for the whole of one's life.

The process of alternately increasing and decreasing the size of the chest is under the control of a collection of nerve cells situated in the brain in the reticular formation of the medulla known as the **respiratory centre.** Nerves run from this centre to the respiratory muscles.

Inspiration is brought about by the contraction of the **intercostal** muscles, which raise the chest wall upwards and outwards, and by the contraction of the **diaphragm** muscle [FIG. 73] which moves downwards and compresses the abdominal contents.

Expiration is brought about by the relaxation of these muscles. The elastic recoil of the lungs and chest wall returns the chest to the resting respiratory level.

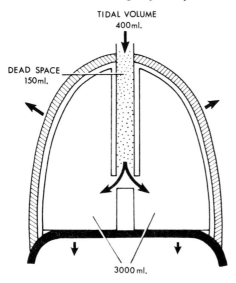

TIDAL VOLUME
400 ml.

DEAD SPACE
150 ml.

3000 ml.

FIG. 73. Lungs contain 3,000 ml. of air at the end of expiration. With next inspiration 400 ml. of air are taken in through nose or mouth. Volume of dead space is 150 ml. and only 250 ml. of room air reach the lungs.
The increase in chest size is brought about by the diaphragm which moves downwards and the contraction of the intercostal muscles which move the chest wall upwards and outwards.

The increase in size of the chest with inspiration is quite small in quiet breathing, and the volume only increases from 3,000 ml. to 3,400 ml. As a result of the increase in volume of 400 ml. this quantity of air enters the nose or mouth.

Dead Space

Only the first 250 ml. of the room air which is taken in with each breath ever reaches the lungs. The reason for this is that the air passages have a volume of 150 ml. and before the last 150 ml. of room air have reached the lungs, the respiratory process is reversed and this air is breathed out again.

The first 150 ml. of each expiration is **dead space air,** which has only reached the air passages. The last 250 ml. of expired air will consist of alveolar air which has come from the lungs, and will contain the carbon dioxide to be excreted from the body. The presence of a **dead space** makes breathing inefficient, because muscular activity is used to take in air which does not reach the lungs.

If 400 people enter a cinema which only has 250 seats, only the first 250 will reach the auditorium whilst the last 150 will have to stay in the corridors outside. At the end of the performance when everybody comes out, only the first 250 in will have seen the film, the remaining 150 will never have reached the important area.

Respiration can be made more efficient in a patient with respiratory distress, by carrying out a **tracheostomy** operation, and opening the trachea to the outside in the region of the neck. Room air can then reach the lungs via a relatively small dead space. So a much smaller tidal volume will suffice to maintain adequate **alveolar ventilation** [p. 48].

In a normal person the large volume of the dead space (150 ml.) means that 150 ml. of each tidal volume is ineffective, and should the patient breathe shallowly, so that the tidal volume is much reduced, the patient will asphyxiate when the tidal volume falls to 150 ml. Any tidal volume of 150 ml. or less will be equivalent to no ventilation at all, since none of the room air will ever reach the lungs. This is an important point to remember when dealing with artificial respiration, because unless the manoeuvre adopted produces a tidal volume that is large compared with 150 ml. it will be completely ineffective.

Underwater swimming using a snorkel tube is an example where the dead space is artificially increased. The air must now be taken in from the end of the snorkel tube, and it has a much larger volume to traverse before it reaches the lungs. In order to breathe through such a tube much deeper breaths are needed. Adequate ventilation of the lungs will be impossible if the tubes are too long and too large in diameter.

Respiratory Muscle Activity

Respiratory movements are brought about by the **respiratory centre** in the medulla [p. 54] using respiratory muscles which are in the category of striated muscles. Striated muscles are the muscles usually employed for voluntary movements. They are supplied by nerves which take their origin in the anterior horn cells of the spinal cord, and the motor nerves pass out via the ventral nerve roots to run to the muscle fibres of the respiratory muscles [see p. 120].

Inspiration is an **active** process, that is, it is brought about by the contraction of the inspiratory muscles. **Expiration** is mainly **passive.** It is brought about by the elastic recoil of the lung and the chest wall. We have **expiratory muscles,** but these are not used except in very severe exercise and when we wish to make a powerful forced expiration such as when blowing up a balloon, or blowing up a football.

When respiration is increased, such as in exercise, **accessory muscles** are employed, in addition to the normal inspiratory muscles. These accessory muscles are employed to raise the thoracic cage, and hence increase the volume in the thorax. The chief accessory muscles which are all attached to the cervical region of the spinal column or the skull, are the sternomastoid, which originates from the sternum and clavicle, and which is inserted into the mastoid process of the skull, the scalene muscles, and trapezius (upper part). The pectorals are also used if the shoulder girdle is fixed by holding on to a rigid object.

Expansion of the Lungs

The lungs are elastic structures and outside the thorax they collapse like a deflated balloon. In the thorax they are fully expanded, and completely fill the thoracic cavity. There is no space between the outside of the lungs and the inside of the thorax.

Should air enter between the outside of the lungs and the inside of the thorax, the lungs will collapse, and this is termed a pneumothorax (L. *pneumo* = air). When a pneumothorax has occurred, respiratory movements no longer produce an adequate ventilation of the lung. A pneumothorax may result from a chest injury, which will allow air into the chest, or it may result from a hole in the lungs, such as from a burst bulla which will allow air between the lungs and the thorax, causing the lung to collapse. The mediastinum separates the two sides of the chest, and as a result a unilateral pneumothorax will only put one lung out of action. Breathing is still possible using the other lung but will be hampered by the mediastinal shift which occurs [see FIG. 74].

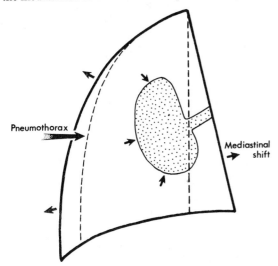

FIG. 74. Pneumothorax and mediastinal shift. Should air enter between lungs and the chest wall the lung collapses whilst the chest wall moves slightly outwards. The mediastinum is pushed over to the other side. The initial position of the chest wall and mediastinum is shown by the dotted lines.

To prevent collapse of the lungs, an underwater seal must be employed with a drainage tube following a thoracotomy operation.

Balloon in the Bottle Demonstration

The way in which the lungs are kept inflated in the thorax may be demonstrated using an ordinary rubber balloon in a glass bottle. The balloon [FIG. 75] is tied on to a piece of glass tubing, so that it can be inflated by blowing down the glass tubing. Normally the only way to keep such a balloon fully inflated, is to tie a piece of string round the neck of the balloon. However, if the balloon is inserted in the bottle and after it has been inflated, the cork is inserted into the side-tube, the

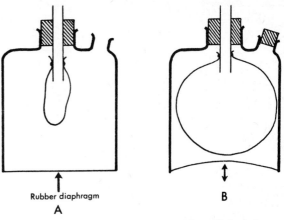

FIG. 75. Balloon in bottle demonstration. If the cork is inserted into the bottle after the balloon has been inflated the rubber diaphragm is displaced upwards but the balloon remains inflated. It is now possible to look down the tubing into the inside of the balloon (B). By moving the rubber diaphragm up and down it is possible to change the air in the balloon. If the cork is removed (A) the balloon collapses and movement of the rubber diaphragm has no effect.

balloon remains inflated and it is now possible to look down the glass tubing into the inside of the balloon. When the cork is removed [FIG. 75(A)] the balloon immediately collapses again. With the balloon in the inflated position, it is possible to change the gas in the balloon by pulling on the rubber diaphragm which forms the base of the bottle [FIG. 75(B)].

LUNG VOLUMES

The volume changes of the lungs with breathing are difficult to assess accurately using a tape measure. All this will give is the increase in the circumference of the chest with breathing, and since this increase is very small, it is difficult to measure accurately. For an accurate determination of the volume changes with breathing a spirometer is used [FIG. 76]. The patient breathes through a tube into the spirometer bell which is forming a water-tight seal to prevent the escape of gas. As the subject breathes in the bell descends, and as the subject breathes out the bell rises. This makes a recording on paper which is calibrated in millilitres of air.

FIGURE 77 shows a typical tracing with quiet breathing, and it shows that the size of the chest increases by 400 ml. with each breath. This is not the maximum inspiration a person can take; if a deeper inspiration is made, the chest volume will increase to a total volume of about 5–6 litres. This increase in inspiration is termed the **inspiratory reserve volume.** It is also possible for the subject to breathe out below the **resting respiratory level,** such as by making a forced expiration. The volume of gas breathed out, this time, is termed the **expiratory reserve volume.** Even when the expiratory reserve volume has been expired there is still a **residual volume** of

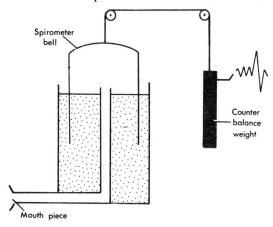

Fig. 76. Spirometer. The subject breathes into the mouth-piece and causes the spirometer bell to move up and down. A writing pen attached to the counter-balance weight enables a permanent record to be obtained.

The volume changes of the spirometer bell are equal to those of the lungs.

gas in the lungs. This is a volume of $1-1\frac{1}{2}$ litres that can never be expired, no matter how great an expiration is made, for the simple reason that it is not possible to approximate the back of the chest to the front of the chest in order to squeeze out this air.

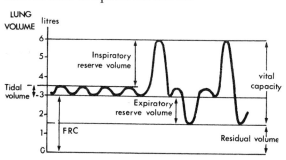

FRC = Functional residual capacity

Fig. 77. Lung volumes.

The volume of the lungs at the resting respiratory level (that is after a quiet expiration and prior to the next inspiration,) is termed the **functional residual capacity** (F.R.C.) It is equal to the expiratory reserve volume plus the residual volume.

Vital Capacity

If the largest possible inspiration is taken and then followed by the largest expiration, the amount of air breathed out is termed the **vital capacity.** It is the largest tidal volume that it is possible to make. Its magnitude

depends on the size of the person, and is usually of the order of $4\frac{1}{2}$ litres in men, and $3\frac{1}{2}$ litres in women.

An important clinical test is the **timed vital capacity** in which the subject is asked to breathe out as quickly as possible from the point of maximum inspiration to the point of maximum expiration. The amount breathed out in the first second should, in a normal person, be at least 80 per cent. of the total vital capacity. This percentage is termed the FEV_1 (which is short for **Forced Expiratory Volume in 1 second**).

In a patient with narrowed air passages, such as a patient with asthma, the amount breathed out in the first second is very much less than this. The effectiveness of a vasodilator drug such as an isoprenaline spray, can be tested using such a technique.

PULMONARY VENTILATION

It has been seen that under resting conditions 400 ml. of air are taken in with each breath, and this is termed the **tidal volume.** Since this process is repeated 15–20 times per minute, the total amount of air taken in per minute will be obtained by multiplying the tidal volume, by the number of times this volume is taken in per minute. This latter figure is termed the respiratory rate.

Thus pulmonary ventilation (V) = tidal volume × respiratory rate.[1]

Numerically, it will range from 400 ml. × 15 = 6,000 ml./minute up to 400 ml. × 20 = 8,000 ml./minute, namely from 6–8 litres/minute.

In exercise the respiratory rate and tidal volume are increased markedly, and the pulmonary ventilation, in this case, may be as high as 50 litres/minute.

This is not the maximum possible ventilation. Voluntarily it is possible to breathe with a very much greater depth, and at a very much faster rate and to take in air in excess of 100 litres/minute. A test which involves breathing maximally in this manner (for, say, 15 seconds) is termed the maximum ventilation volume (maximum breathing capacity).

Alveolar Ventilation

The quantity of room air that reaches the lungs per minute is less than the pulmonary ventilation because the pulmonary ventilation includes air that only reaches the dead space.

$$\text{Alveolar ventilation} = \text{Respiratory rate} \times (\text{tidal volume} - \text{dead space})$$
$$= 15 \times (400 - 150)$$
$$= 3,750 \text{ ml./minute}$$

(Pulmonary ventilation = 6,000 ml./minute.)

The alveolar ventilation also gives the quantity of alveolar air breathed out into the room per minute.

[1] V = volume. V̇ = volume per minute.

Since the alveolar air contains a constant 5·5 to 6 per cent. CO_2, the quantity of CO_2 given off per minute will be directly proportional to the alveolar ventilation.

EXAMPLE:

$$CO_2 \text{ given off per minute} = \text{alveolar ventilation} \times \\ \%CO_2 \text{ in alveolar air} \\ = 3{,}750 \times 5\cdot5\% \\ = 3{,}750 \times \frac{5\cdot5}{100} \\ = 206 \text{ ml. } CO_2 \text{ per minute.}$$

Summary

1. At the end of a quiet expiration the LUNGS contain 3 litres of ALVEOLAR AIR (6% CO_2, 14% O_2). The blood flowing through the lungs takes up oxygen from (and gives off CO_2 to) this alveolar air.

2. The body uses 250 ml. O_2/minute at rest (up to 5 litres/minute during severe exercise) and gives off 200 ml. CO_2 per minute.

3. O_2 is replaced by inspiring ROOM AIR (0% CO_2, 21% O_2) by increasing the size of the chest 15 to 20 times/minute (RESPIRATORY RATE).

4. Chest volume is increased by 400 ml. and this volume of air enters via nose or mouth (TIDAL VOLUME).

5. PULMONARY VENTILATION = 6 litres/min.

RESPIRATORY RATE × TIDAL VOLUME
15 400 ml.

6. But 150 ml. DEAD SPACE ∴ only 250 ml. room air reaches lungs.

7. Increase in size of chest brought about by contraction of RESPIRATORY MUSCLES controlled by RESPIRATORY CENTRE in MEDULLA.

9. TRANSPORT OF OXYGEN AND CARBON DIOXIDE IN THE BLOOD

CARRIAGE OF GASES BY THE BLOOD

When considering the carriage of a gas such as oxygen in the blood, it is necessary to consider two entirely different factors:

(*a*) tension
(*b*) quantity.

Tension

Tension is the driving force which drives a gas from one region to another. Gases always move from a region of high tension to a region of low tension. In this way they resemble electricity which passes from a region of high tension or voltage to a region of low tension.

Oxygen, for example, in the lungs, passes from alveolar air into the blood in the pulmonary capillaries because the tension of oxygen in the alveolar air is higher than in the blood. When the blood arrives at the tissues the oxygen passes from the blood to the tissues because now the tension of oxygen in the blood is higher than that in the tissues.

Quantity

Continuing with the electrical analogy, when buying a dry cell battery for a flash-lamp or transistor radio, it is possible to buy a physically large battery or a physically small battery both of which will give the same tension or voltage. The difference between the large battery and the small battery is that the large battery holds a greater quantity of electricity than the small battery.

When considering the oxygen in the blood, it is important to know not only the tension or driving force, but also the quantity, i.e. how much of it is present. In the case of oxygen, this will depend on the amount of haemoglobin in the blood. A person who is anaemic may have the same oxygen tension in the arterial blood as a normal person, but will have a very much lower oxygen content because of the deficiency of haemoglobin.

GAS TENSIONS

The tension (partial pressure) of oxygen in the alveolar air of the lungs depends on the barometric pressure, and on the tension exerted by the other gases which are present.

In the lungs there are four gases:

oxygen
carbon dioxide
nitrogen
water vapour

and these four gases together make up the total barometric pressure of 760 mm. Hg.

The water vapour which is fully saturated in the lungs, exerts a tension of 47 mm. Hg. This is a physical property of water which depends only on temperature and is independent of the presence of the other gases. The tension exerted by water (water vapour pressure) increases with temperature and at a 100°C. the tension of water has risen to 760 mm. Hg and then equals the barometric pressure. A liquid boils when its tension equals the barometric pressure. This is the reason why water boils at 100°C. At the top of a mountain where barometric pressure is only 400 mm. Hg, water would boil at 80°C. It is not possible to brew tea satisfactorily at this altitude since the water temperature of boiling water is not high enough.

It is possible to make water boil at a temperature higher than 100°C. by increasing the barometric pressure. This principle is used in the pressure-cooker, and hospital **sterilizers.** The barometric pressure is increased by using a closed container and the pressure is allowed to increase sufficiently to raise the boiling point of water to 115°C. or more. This higher temperature is needed in sterilization to destroy the spores of organisms which would not be destroyed at 100°C.

Returning to a consideration of gas tensions in the lungs, we have seen that the water vapour exerts a tension of 47 mm. Hg and thus contributes a pressure of 47 mm. Hg towards the total pressure of 760 mm. Hg. It follows that the remaining pressure of 713 mm. Hg is due to the presence of the other three gases, oxygen, carbon dioxide and nitrogen. Each of these gases contributes to the total pressure, in proportion to its percentage in the mixture.

Thus oxygen contributes 14 per cent. of the total of 713 mm. Hg, carbon dioxide contributes 6 per cent., whilst nitrogen contributes the remainder.

14 per cent. of 713 mm. Hg is 100 mm. Hg and this is the tension of oxygen in the lungs. It is denoted by P_{O_2}. or pO_2.

The contribution of carbon dioxide towards the total pressure (tension of carbon dioxide) is 6 per cent. of 713 mm. Hg which equals 40 mm. Hg. This is the tension of carbon dioxide in the lungs [TABLE 1].

TABLE I

ALVEOLAR AIR IN LUNGS

Composition	Tensions
14% oxygen	P_{O_2} = 100 mm. Hg
6% carbon dioxide	P_{CO_2} = 40 mm. Hg
80% nitrogen	P_{N_2} = 573 mm. Hg
Water vapour (saturated)	P_{H_2O} = 47 mm. Hg
Total	= 760 mm. Hg

CARRIAGE OF OXYGEN

When a liquid such as water (or blood) is exposed to air the tension of oxygen in the air drives oxygen into the water [FIG. 78]. As more and more oxygen enters

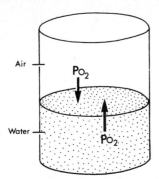

FIG. 78. If water is exposed to air the gases in the air come into equilibrium with those in the water. This partial pressure of oxygen in the air, Po2, will be equal to the tension of oxygen in the water, Po2.

the lung the tension builds up and oxygen tries to leave. An equilibrium is set up and no further net exchange occurs when the tension of oxygen in the water (Po2) is equal to the tension of oxygen in the air.

OXYGEN TENSION

The blood as it passes through the lungs takes up oxygen because the tension of oxygen in the lungs is higher than the tension of oxygen arriving via the blood. The blood leaves the lungs at a tension of 100 mm. Hg and passes via the pulmonary veins to the left side of the heart [FIG. 79]. It then passes via the aorta and arteries to the capillaries of the tissues, but no change in oxygen tension occurs until the tissues are reached. The blood thus arrives at the tissue capillaries with the same tension of 100 mm. Hg.

As the blood flows through the tissue capillary, it comes in contact with the tissue fluid which has a much lower oxygen tension. A typical figure is 40 mm. Hg. The tension of oxygen in the tissue fluid is low because oxygen is continually being used up by the process of metabolism.

As the blood flows through the capillary the oxygen tension falls from 100 to 90 to 80 to 70 to 60 to 50 to 40 mm. Hg, and then it is at the same tension as the surrounding tissue fluid. No further oxygen is given up, and for the remainder of its passage through the capillary there is no further interchange.

The blood returns to the right side of the heart via the veins, with an oxygen tension of 40 mm. Hg. It passes through the right side of the heart, and reaches the lungs via the pulmonary artery. It arrives in the pulmonary capillaries with a tension of 40 mm. Hg, and comes in contact with the oxygen in the alveolar air at a tension of 100 mm. Hg. The tension of oxygen in the lung capillary blood builds up, from 40 to 50 to 60 to 70 to 80 to 90 to 100 mm. Hg. It is then in equilibrium

with the alveolar air, and it leaves the lungs via the pulmonary veins with this tension.

Thus the oxygen is moving, all the time, from a region of high tension to a region of low tension [FIG. 79].

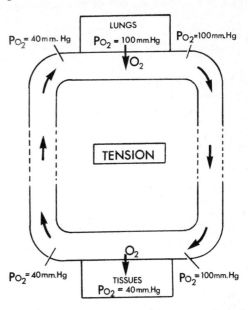

FIG. 79. Tension changes involved in the carriage of oxygen.

Oxygen Electrodes

The tension of oxygen in blood may be measured directly by using an oxygen electrode. The principle on which this measuring device is based, is that when a small voltage (0·6 volt) is applied between a platinum and a silver electrode in a fluid, the electric current which flows depends on the tension of oxygen. Since the proteins in blood would 'poison' the platinum electrode, the blood is separated from the platinum electrode by means of a thin plastic membrane which will allow oxygen to pass through but prevents the passage of plasma proteins.

The blood sample is drawn from the blood vessel using a syringe and introduced into the oxygen electrode. Special care must be taken to ensure that the blood sample is not exposed to room air as this would alter the oxygen tension. The oxygen tension is 'displayed' on a meter.

OXYGEN CONTENT

The quantity of oxygen carried depends on the property of haemoglobin to combine with oxygen. One gram of haemoglobin has the ability to combine with 1·34 ml. of oxygen.

Thus a person who has 15 grams of haemoglobin in every 100 ml. of blood, would, theoretically, be able to carry $1·34 \times 15 = 20$ ml. of oxygen in every 100 ml. of blood. This is termed the **oxygen capacity** of the blood.

It is the amount of oxygen carried by the blood when the tension is very high and the haemoglobin is fully saturated with oxygen.

At a tension of 100 mm. Hg the amount of oxygen carried by the blood is close to this maximum. It is in the region of 95–97 per cent. of the oxygen capacity. Thus, instead of 20 ml. of oxygen being carried by the blood, only 19 ml. of oxygen are present in every 100 ml. of blood leaving the lungs via the pulmonary veins, and reaching the arteries of the body. This is termed the **arterial oxygen content**.

In **venous blood**, that is, the blood returning from the tissues via the veins to the right side of the heart, only 14 ml. of oxygen are present in every 100 ml. of blood [FIG. 80] and this is the oxygen content of the blood

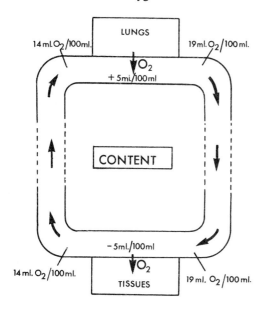

FIG. 80. Content changes associated with the carriage of oxygen.

arriving at the lungs under resting conditions. This blood still contains an appreciable amount of oxygen. The tension has fallen to 40 mm. Hg, but the content has only fallen to 70 per cent. of the oxygen capacity, i.e. 14 ml. of oxygen per 100 ml. blood. This is because the curve connecting the quantity of oxygen carried and the oxygen tension is not a straight line, but an S-shaped curve known as the **oxygen dissociation curve** [FIG. 81].

The curve shows that the blood leaving the lungs has a tension of 100 mm. Hg and a content of 19 ml. of oxygen per 100 ml. of blood at point A. It arrives at the tissues with the content and tension unchanged, but as the blood passes through the tissue capillary it moves from point A to point B and leaves the tissues with a tension of 40 mm. Hg and a content of 14 ml./100 ml. of blood. Every 100 ml. blood which flows through the tissues thus gives up 5 ml. oxygen. In exercise more is given up.

Such a dissociation curve with oxygen content expressed in ml. oxygen/100 ml. blood would only apply to a person who has 15 grams of haemoglobin/100 ml. blood. The curve could be made more universal by plotting along the vertical axis the percentage saturation, i.e. from 0 to 100 per cent. Such a curve could then be used for an anaemic person by replacing the 100 per cent. with the oxygen capacity of that patient.

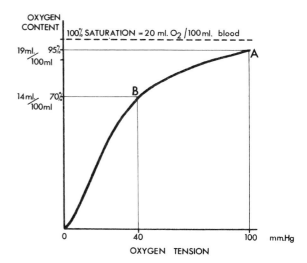

FIG. 81. Oxygen dissociation curve. At oxygen tensions higher than 100 mm. Hg (95% saturation) there will only be a slight increase in oxygen content. Above 150 mm. Hg oxygen tension the haemoglobin is fully saturated with oxygen (100% saturation).

If, for example, the person had only half the normal haemoglobin, the 100 per cent. would then correspond, not to 20 ml. oxygen per 100 ml. blood, but to 10 ml. oxygen per 100 ml. blood.

High Oxygen Tensions

The oxygen tension in the lungs depends on the barometric pressure, and at sea-level may be obtained approximately by multiplying the percentage of the oxygen in the alveolar air by 7. The normal oxygen tension in the lungs is 100 mm. Hg but this may be increased by breathing an oxygen-enriched mixture. In this way, the oxygen tension in the lungs can be increased to several hundred mm. Hg. Breathing pure oxygen, the tension approaches 700 mm. Hg.

As will be seen later, such an increase in oxygen tension does not necessarily mean there is an increase in the quantity of oxygen carried by the blood.

Low Oxygen Tensions

At high altitudes the total barometric pressure will be low, and as a result, even though the percentage of oxygen in the alveolar air is normal, namely 14 per cent., oxygen tension will be low, and the subject will suffer from oxygen lack.

CARRIAGE OF CARBON DIOXIDE

Carbon dioxide is a gas that dissolves in water and forms carbonic acid.

$$CO_2 + H_2O \rightarrow H_2CO_3$$
carbon dioxide   water   carbonic acid

CARBON DIOXIDE TENSION

We have seen that the tension of carbon dioxide in the lungs is 40 mm. Hg. The tension of carbon dioxide in the tissues is 46 mm. Hg and as the blood flows through the tissue capillaries, the tension of carbon dioxide in the blood builds up to 46 mm. Hg. The blood leaves the tissue capillaries and returns via the veins to the right side of the heart and then to the pulmonary artery and the lungs with the carbon dioxide at a tension of 46 mm. Hg [FIG. 82]. As it passes through the lung

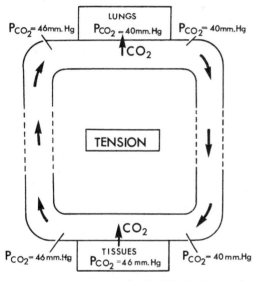

FIG. 82. Tension changes involved in the changes in carbon dioxide.

capillaries, carbon dioxide passes from the blood into the alveoli of the lungs because the tension in the blood is higher than that in the lungs. Thus the tension falls from 46 to 45 to 44 to 43 to 42 to 41 to 40 mm. Hg. It is then in equilibrium and no further fall in carbon dioxide tension occurs.

The blood leaves the lungs via the pulmonary veins to the left side of the heart and then to the arteries of the body with a tension of 40 mm. Hg. It arrives at the capillaries with this tension and the tension builds up from 40 to 41 to 42 to 43 to 44 to 45 to 46 mm. Hg as carbon dioxide moves from the tissue fluid into the blood. We have completed the circle and the carbon dioxide leaves the tissues at a tension of 46 mm. Hg.

It will be noted that carbon dioxide moves in the opposite direction to oxygen because the tension of carbon dioxide in the tissues is higher than that in the lungs.

CARBON DIOXIDE CONTENT

The carbon dioxide content of the blood leaving the lungs is 48 ml. carbon dioxide per 100 ml. blood [FIG. 83]. The blood arriving at the tissues will have the same content namely 48 ml. CO_2/100 ml. blood. As it passes through the tissue capillaries, the content increases from 48 ml. to 52 ml. of CO_2/100 ml. blood. There is thus

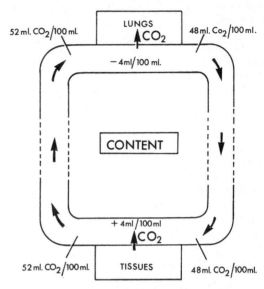

FIG. 83. Content changes associated with carriage of carbon dioxide.

an uptake of 4 ml. CO_2 for every 100 ml. of blood which flows through the tissues. The content remains unchanged until the lung capillaries are reached, when the content falls from 52 to 48 ml. of CO_2 per 100 ml. blood as the blood passes through the lungs.

There are two surprising points about the carriage of carbon dioxide. The first is that the blood contains very much more carbon dioxide per 100 ml. than it does oxygen. The arterial blood contains 48 ml. of CO_2, but only 19 ml. oxygen per 100 ml. blood. The second point is that as the blood passes through the lungs, it only gives up a relatively small amount of the carbon dioxide. The blood arrives at the lungs with 52 ml. of CO_2 per 100 ml. blood and even after the blood has passed through the lungs, it still contains 48 ml. of CO_2. The reason for this is that carbon dioxide is more than a waste product. An adequate level must be maintained in the blood for the proper functioning of the body.

BLOOD pH

Respiration and the carriage of carbon dioxide are closely interrelated with the maintenance of the blood at the correct pH level.

The pH scale is a notation that extends from 0 to 14 with 7 as its middle point [FIG. 84]. It is a scale that measures **hydrogen ion concentration.**

Normal water ionizes very slightly into hydrogen ions and hydroxyl ions. There are so few hydrogen ions present in pure water, that 10,000,000 litres (10^7 litres) are needed to provide one gram of hydrogen ions. Such a solution, which is neutral, has a pH of 7.

Alkaline solutions have much fewer hydrogen ions than pure water. A molar sodium hydroxide solution has so few hydrogen ions that 10^{14} litres of this solution, that is 100,000,000,000,000 litres, would be required to provide one gram of hydrogen ions. Such a solution has a pH of 14.

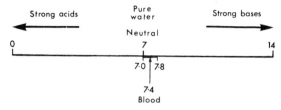

FIG. 84. pH notation. Water is neutral (pH 7). Acids have a pH of less than 7. Bases (or alkalis) have a pH greater than 7.
Blood alkaline at pH 7·4 (range for life 7·0–7·8).

Acid solutions have more hydrogen ions than pure water, and only 10 litres of hydrochloric acid of the same strength $\left(\dfrac{M}{10}\right)$ as released by the gastric cells of the stomach, are required to provide one gram of hydrogen ions. This solution has a pH of 1.

The pH scale of hydrogen ion concentration is, thus, a scale in which the pH is equal to the number of 0's following 1, of the litres required to supply one gram of hydrogen ions. So if 10 litres are required, the pH is 1, if 100 litres are required the pH is 2, 1,000 litres required the pH is 3, 10,000,000—pH is 7, and 100,000,000—pH is 8.

Mathematically, the pH is the negative logarithm of the hydrogen ion concentration, or the power to which 10 must be raised to give this concentration with its sign changed.

Referring to FIGURE 84 any solution which has a pH of less than 7 is acid, and any solution which has a pH of over 7 is alkaline.

Blood is alkaline since its pH is 7·4. The blood pH must be kept between 7·0 and 7·8 for life to continue. The reason why it is so important to maintain the blood at the correct pH is that the behaviour of proteins varies very greatly with pH, and all the enzymes in the body are proteins.

Carbon Dioxide and Blood pH

Carbon dioxide is carried in the blood in three ways.
Firstly, in simple solution, and forming carbonic acid.
Secondly, as sodium bicarbonate in the plasma, and potassium bicarbonate in the red cells.
Thirdly, as neutral carbamino protein, mainly with the haemoglobin in the red cells.

The transport of the acid-gas carbon dioxide in the blood is closely connected with the maintenance of blood pH at 7·4. Carbon dioxide forms the relatively strong acid, carbonic acid, which would on its own have a pH of around 4. In the plasma, the carbon dioxide in this form is neutralized by carbon dioxide in the form of sodium bicarbonate. In order to keep the pH at 7·4, a 20:1 ratio must be maintained between the carbon dioxide which is in the form of bicarbonate, and the carbon dioxide in solution which forms carbonic acid.

The bicarbonate is manufactured in the red blood cell. Thus when the carbon dioxide is picked up as the blood passes through the tissue capillaries, most of the carbon dioxide passes through the plasma to the red cell, and in the red cell the carbon dioxide is rapidly converted to carbonic acid by an enzyme in the red cell called **carbonic anhydrase.** The carbonic acid reacts with the haemoglobin which is giving up its oxygen, and forms bicarbonate.

$$H_2CO_3 \rightarrow H^+ + HCO_3^-$$
carbonic acid   hydrogen ions   bicarbonate

The hydrogen ions combine with the haemoglobin leaving the negatively charged bicarbonate ions. Since the main positively charged ions in the red blood cell are potassium, this is equivalent to the formation of potassium bicarbonate.

Most of the bicarbonate which is manufactured in this way in the red cell, passes out into the plasma, in exchange for chloride, which moves into the red cell. When the bicarbonate enters the plasma, where the principal positively charged ion is sodium, it becomes sodium bicarbonate. The chloride entering the red cell becomes potassium chloride.

The whole process is reversed as the blood goes through the lungs, the chloride comes out of the red cell into the plasma, the bicarbonate goes from the plasma into the red cell where it is broken down by the carbonic anhydrase to carbon dioxide and water, which then passes out to the plasma and through the alveolar membrane into the alveoli of the lungs. The movement of chloride into the red cells as the blood passes through the tissues, and out of the red cells in the lungs is termed the **chloride shift.**

CONTROL OF RESPIRATION

The respiratory centre is under the influence of a number of factors [FIG. 85], the most important of which are the **blood pH** and the level of **carbon dioxide** in the blood. Should the level of carbon dioxide in the blood be too high, the respiratory centre is stimulated and respiration is increased. The effects of the increased respiration will be to make the alveolar air more like room air, that is, the alveolar carbon dioxide will be reduced from 6 per cent. towards 0 per cent. As the carbon dioxide is washed out of the lungs, so the tension of carbon dioxide in the arterial blood falls, and a new equilibrium will be set up with the carbon dioxide in the arterial blood at the correct level.

Conversely if the carbon dioxide level in the blood is too low, there is less drive on to the respiratory centre, and respiration is temporarily depressed, until the carbon dioxide in the blood has built back to its original level.

Oxygen shortage stimulates breathing via the **chemoreceptors.** These are cells found in the **carotid body** and **aortic body,** which sample blood for oxygen content. When the oxygen content falls, they send nerve impulses to the respiratory centre to stimulate breathing.

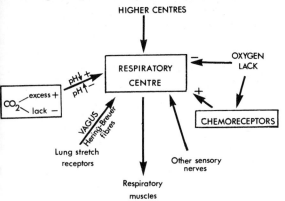

FIG. 85. Factors controlling respiration.

If the **oxygen lack** is very severe, it **depresses** the **respiratory centre** by a direct action, and the chemoreceptor stimulation is no longer effective.

The **higher centres** modify the respiratory centre activity. For a time it is possible to take over breathing from the respiratory centre, and to voluntarily over- or under-breathe. **Emotional excitement** stimulates breathing via the respiratory centre and if severe, may lead to a washing out of carbon dioxide with serious effects.

The higher centres modify the breathing pattern completely when **talking,** singing, etc. because in these cases the breathing pattern is determined by the sentences spoken or the musical passage being sung. It is very difficult to speak when breathing in, and speech and singing usually only occur during expiration. The respiratory pattern when carrying out these manoeuvres consists of a very slow prolonged expiration followed by a very short sharp inspiration.

Breathing stops when **swallowing.** If it does not then the food goes down the wrong way, that is, it is inhaled into the larynx and trachea instead of passing down the oesophagus. Respiration is halted at any phase of the respiratory cycle by the thoracic cage muscles and the diaphragm stopping contraction instantly, and holding the chest at a certain volume until the swallowing has occurred.

Swallowing is no longer possible if the subject is **unconscious,** so no fluids should ever be given to an unconscious person because they will be inhaled into the lungs with the next inspiration. When unconscious the tongue may flop back and obstruct the airway. If this occurs breathing will cease and it is necessary to extend the head, or to move the lower jaw forward to remove the obstruction. A patient returning from the operating theatre after an operation often has an *airway* in position which hooks round the back of the tongue and allows a clear passage of air. Such an airway should not be removed until full consciousness has been regained.

Normal inspiration is limited by the Hering-Breuer reflex. As one breathes in, the lungs expand, and **stretch receptors in the lungs** send sensory information up the **vagus nerve** to the respiratory centre cutting short inspiration. If it were not for these Hering-Breuer impulses, respiration would be much deeper than it is, although the pulmonary ventilation would be unaffected.

MAINTENANCE OF BLOOD pH BY RESPIRATION

The respiratory centre plays a very important part in maintaining the pH of the blood at 7·4. It does so in association with the kidneys, but is able to act more rapidly than the kidneys. It has already been seen that the maintenance of the blood pH at 7·4 depends upon the sodium bicarbonate in the plasma having a concentration of 20 times that of the carbonic acid in the plasma.

If a foreign acid enters the blood, such as ingested sulphuric or phosphoric acid from cheap lemonade, or acetoacetic acid in the case of a diabetic [see p. 69], then the sodium bicarbonate is available to neutralize the acid.

$$\text{Sodium bicarbonate} + \text{sulphuric acid} \rightarrow \text{sodium sulphate (neutral)} + \text{carbon dioxide} + \text{water.}$$

The carbon dioxide which is evolved is given off when the blood next passes through the lungs. The net result, however, is that the plasma sodium bicarbonate is reduced and as a result the ratio of sodium bicarbonate to carbonic acid is now too low, and the blood becomes more acid, that is, its pH falls from 7·4 to 7·3 or lower. This fall in pH stimulates respiration by acting on the respiratory centre, and, as a result, breathing increases. The effect of this is to wash out carbon dioxide from the lungs and to reduce the CO_2 in the blood. The effect is mainly on the carbonic acid and as a result the ratio is restored to 20:1. The over-all effect will be that the subject will have a reduced plasma bicarbonate and will be breathing more deeply and more rapidly, but will be maintaining the blood at its alkaline state of 7·4. This accounts for the clinical condition of air hunger which is seen in patients with untreated diabetes mellitus who have ketone bodies (acetoacetic acid, etc.) in their blood.

In the case of alkalaemia, such as follows the ingestion of large amounts of sodium bicarbonate, the plasma bicarbonate level is too high, and so also is the ratio of sodium bicarbonate to carbonic acid. The effect of this is to depress respiration so that carbon dioxide and hence carbonic acid accumulates in the blood, and once again the pH is restored to normal but with respiration depressed.

ASPHYXIA, HYPERCAPNIA AND ANOXIA

Asphyxia is a name given to the state when there is oxygen lack and carbon dioxide excess in the body. Such a condition can arise when respiration is insufficient for the body's needs, when there is respiratory obstruction, or when one is breathing in a confined space so that the expired air is re-inhaled.

Asphyxia produces a marked stimulation in respiration, and this is due to the combined effects of the carbon dioxide excess acting directly on the respiratory centre, and the oxygen lack which is acting via the chemoreceptors. Since carbon dioxide is being retained, too much carbonic acid will be present in the body and the blood will become slightly acid.

The state when there is too much carbon dioxide only in the body is called **hypercapnia** or hypercarbia. A carbon dioxide excess without a concurrent oxygen lack is an unusual event and only arises when breathing a gas mixture which contains a high percentage of carbon dioxide, such as 5 per cent. or 7 per cent., and adequate oxygen. Hypercapnia stimulates breathing.

A shortage of oxygen alone, without a concurrent shortage of carbon dioxide, is termed **anoxia,** or, if less severe, **hypoxia.**

Although anoxia in the initial stages may produce a stimulation of respiration via the chemoreceptors, when it is severe, respiration ceases.

Anoxia depresses all parts of the brain, and particularly the higher centres associated with a realization of danger. A person suffering from anoxia becomes disorientated, loses all sense of danger, and passes rapidly into a coma, often without any marked respiratory stimulation. Anoxia is an extremely dangerous state since a person suffering from anoxia loses the will to save himself.

Anoxia has been divided into four different types.

Anoxic anoxia or hypoxic hypoxia is anoxia due to a shortage of oxygen in the inspired air, or due to lung disease so that the oxygen cannot enter the blood. It is associated with a **low tension of oxygen in the arterial blood.** Such a state of affairs may be brought about by breathing a gas mixture which contains insufficient oxygen. It also develops when there is insufficient partial pressure of oxygen such as occurs at high altitudes. Severe anoxic anoxia would be produced if one entered an atmosphere without oxygen. There would be no warning symptoms and one would black out without prior warning. Respiration would cease and shortly afterwards the heart would stop beating.

It should be remembered that if carbon dioxide is also accumulating, the state of **asphyxia** occurs and in this state the subject fights for breath using practically all the body muscles to do so. But in anoxia no such respiratory stimulation occurs.

The second type of anoxia is **anaemic anoxia** which is due to a **deficiency of haemoglobin** in the blood. The oxygen tensions will be normal, but there will be a shortage of oxygen carried by the blood due to the shortage of haemoglobin. In addition to being caused by anaemia, this type of anoxia can also be caused by **carbon monoxide poisoning.** Carbon monoxide is a gas which has a great affinity for haemoglobin. By forming carboxyhaemoglobin it makes the haemoglobin no longer available for the carriage of oxygen. Carbon monoxide is found in coal-gas and in the exhaust fumes of the motor car.

The colour of carboxyhaemoglobin is cherry pink, and a person suffering from this type of anoxia will have an unnaturally pink complexion.

The carbon monoxide has such a strong affinity for the haemoglobin that it may be necessary to subject the patient to an oxygen pressure of several atmospheres (hyperbaric oxygen) in order to remove the carbon monoxide rapidly. In severe cases an exchange blood transfusion is carried out with blood which does not contain any carbon monoxide.

The third type of anoxia is **stagnant anoxia.** This is due to the **blood flowing too slowly** round the circulation, so that although the oxygen tension and oxygen content leaving the lungs are normal, the blood flows so slowly that insufficient oxygen is supplied to the tissues in a given time.

The fourth type of anoxia is **histotoxic anoxia,** and this is due to the failure of the cells to extract oxygen from the blood. This occurs in **cyanide poisoning.**

The various types of anoxia have been likened by J. Barcroft (*Lancet*, 1920) to the supply of milk from a dairy. There may be a shortage of milk supplied to one's house for a number of reasons.

There may be a shortage of milk at the dairy. (This would correspond to a shortage of oxygen—anoxic anoxia.)

There may be a shortage of milkmen to carry the milk. (This corresponds to a shortage of haemoglobin in the state of anaemic anoxia.)

There may be sufficient milk at the dairy, and there may be sufficient milkmen, but the milkmen may go round so slowly that there are long intervals between successive deliveries, and during these intervals a shortage of milk develops. (This corresponds to stagnant anoxia where the blood flow is too sluggish to maintain an adequate oxygen supply to the body.)

Finally there may be a form of milk shortage due to the inability to open the front door[2], thus there may be plenty of milk in the dairy, plenty of milkmen to bring the milk round, the milk arrives but there is still a shortage of milk because it is not possible to take it in. This corresponds to histotoxic anoxia. In this last form of anoxia the tissues are unable to remove the oxygen from the blood and the blood returns to the lungs with such a high oxygen tension and content that even the venous blood is bright red in colour.

OXYGEN EXCESS

The tension of oxygen in the blood may be increased by breathing pure oxygen, and may be increased still further by breathing pure oxygen under a pressure in

---

[2] This analogy has been added by the author since histotoxic anoxia was not recognized as a type of anoxia in Barcroft's time.

excess of one atmosphere. But the oxygen content of the blood will not be increased to any great extent, because the haemoglobin is very nearly saturated when breathing room air at sea level. There will be a small increase, however, in the oxygen dissolved in the plasma. Oxygen under a high pressure is a poisonous gas, and produces irritation of the lungs and toxic effects on the enzymes of the body. Care must be taken, not to subject an individual to too high an oxygen tension for too long a period of time. In the case of premature babies, a high oxygen tension has been shown to produce fibrosis behind the lens of the eye (retrolental fibroplasia) which leads to blindness.

The mental and physical effects on a normal person of breathing pure oxygen at one atmosphere for a short time are nil. There is no detectable change in mental activity, the gas smells the same and tastes the same as room air. There is, however, a slight reduction in heart rate in most people.

Hyperbaric Oxygen

Pure oxygen under a barometric pressure of several atmospheres is termed **hyperbaric oxygen.** To achieve these high tensions a pressure chamber is employed and oxygen tensions well in excess of 1,000 mm. Hg are achieved. Successes are claimed using this technique in treatment of new-born infants who do not breathe, in the treatment of coronary thrombosis in adults, etc. The time that can be spent in such an environment is limited by oxygen toxicity.

Oxygen Toxicity Limits

Oxygen breathed for long periods of time at high pressures produces toxic effects. The limits of human tolerance are not accurately known, but it appears as if oxygen tensions up to 400 mm. Hg are safe.

Exposure to oxygen tensions of two atmospheres (1,500 mm. Hg) produces changes in the lungs and brain after eight hours.

CARBON DIOXIDE LACK

The carbon dioxide in the blood can be reduced by voluntary hyperventilation, but it should be pointed out that this is a potentially dangerous procedure unless carefully supervised. If one breathes deeply and rapidly through the mouth for a period of two minutes, the carbon dioxide tension in the blood can be reduced to about one half of its normal value. The effect of this washing out of the acid gas, carbon dioxide, is to produce alkalaemia of the blood, and the pH rises from 7·4 to 7·5 or higher.

The effect of washing out the carbon dioxide is to increase the excitability of the nerves and muscles of the body, particularly those of the arms and legs. The increased excitability shows itself in the fact that slight pressure on the nerves will bring about muscular contraction, and ultimately the muscles will go into spontaneous spasm, known as carpopedal spasm. (Carpo = wrist, pedal = feet, i.e. spasm of the hands and feet.) This condition is known as **tetany.**

Before tetany sets in, it is possible to demonstrate the increased excitability of nerves by tapping over the angle of the jaw so as to stimulate the facial nerve. Normally this has no effect, but in **latent tetany,** the facial muscles on that side contract. If an arm cuff is placed around the arm and inflated, this will precipitate the spasm of the wrist and hand. (See also tetany associated with underactivity of the parathyroid gland, p. 103.

At the same time the loss of carbon dioxide, which is a vasodilator to the blood vessels of the brain, brings about a reduction in cerebral blood flow. Vision is impaired, and the peripheral fields become narrowed. Clear thought becomes impossible, and the subject may complain of dizziness.

It is important to remember these symptoms since hyperventilation may result from emotional excitement, and one should be on one's guard against any patient who is sitting up in bed over-breathing. The treatment of carbon dioxide lack due to hyperventilation is to get the subject to breathe a gas mixture containing carbon dioxide. If such a gas mixture is not readily available, as a first-aid measure, the subject may be made to rebreathe his own expired air from a paper bag. An hysterical state may develop following hyperventilation which results in the hyperventilation continuing without the subject being aware that this is going on. When the carbon dioxide level to the brain has been increased this state gradually declines and returns to normal.

CYANOSIS

A patient suffering from anoxia becomes visibly **cyanosed** (Greek *Kyanos* = blue) when the oxygen content of the skin blood vessels is so low that more than five grams of haemoglobin per 100 ml. blood is in the reduced form (that is not combined with oxygen). **Cyanosis** is seen in patients with **anoxic anoxia** and **stagnant anoxia.** It should be remembered, however, that a severely anaemic patient may have insufficient haemoglobin to show cyanosis no matter how severe the anoxic anoxia.

10. DISORDERS OF RESPIRATORY FUNCTION

DISORDERS WHICH AFFECT THE MECHANICAL PROCESS OF BREATHING

Some of the commoner pathological changes which occur in the respiratory system are as follows:

1. Inspiratory Failure

As has been seen, inspiration is brought about by the activity of the respiratory centre which sends nerve impulses to the anterior horn cells supplying the respiratory muscles. The anterior horn cells relay these nerve impulses to the muscles themselves.

The **respiratory centre** is part of the brain and may be damaged by any form of injury which affects the brain, e.g. trauma, anoxia or excessive doses of drugs including anaesthetics. If the respiratory centre stops functioning, then natural respiration will cease. Artificial respiration will then be required to maintain life.

Poliomyelitis is a virus infection which attacks the anterior horn cells of the spinal cord. Should the segments of the spinal cord controlling the respiratory muscles be affected by this disease, then respiration will be impaired by the paralysis of these respiratory muscles. If this loss of muscle power is extensive, artificial respiration will be required until, if ever, these muscles recover. An iron-lung (cabinet respirator), or a positive pressure respirator attached to a tracheostomy tube, is employed.

The American-Indian arrow poison, curare, brings about paralysis by blocking the transmission of nerve impulses at the motor end-plates associated with the voluntary muscle fibres [p. 124]. The derivative tubocurarine is used as a muscle relaxant. When such a muscle relaxant is employed to reduce the level of anaesthesia required to bring about relaxation of, say, the abdominal muscles, it will also tend to paralyse the respiratory muscles. In such a case artificial respiration must be employed for the duration of the operation and continued until the effect of the muscle-relaxant drug has disappeared.

2. Expiratory Failure

Since expiration is brought about by the elastic recoil of the lungs and only in exceptional cases by expiratory muscles, expiratory failure is due to a loss of elasticity. It results from the breaking down of the elastic tissue in the lungs as the consequence of a large number of years of coughing with chronic bronchitis. The effect of such a loss of elastic tissue will be for the chest to move toward the full inspiratory position. This gives the typical barrel-shaped chest of emphysema which is associated with difficulty in expiration.

3. Pneumothorax

It has been seen [p. 47] that if air enters between the lungs and the chest wall, then respiration is impaired. Such a pneumothorax is also associated with the shift of the mediastinum to the other side.

4. Pleural Effusion

If fluid enters between the lungs and the chest wall as a result of pleurisy, or if blood enters as a result of an injury (haemothorax) then respiration will be impaired, and once again the mediastinum will be displaced to the other side.

5. Blocked Bronchus

Should a bronchus become blocked so that air can no longer enter a lobe of the lung, then the air that is trapped becomes absorbed into the blood and the lung collapses. The effect, this time, is for the collapsed lung to draw the mediastinum over to its own side.

6. Partially Blocked Airway

In asthma and other conditions of bronchospasm, the air passages become narrowed, so that, although breathing may be satisfactory when the respiratory rate is low, any attempt at exercise causes great embarrassment. The air passages have smooth muscle in their walls

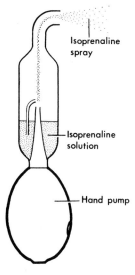

Fig. 86. Bronchodilator spray. Squeezing the hand-pump causes a fine spray of isoprenaline or salbutamol which is inhaled. This causes dilatation of the smooth muscle of the respiratory system and relieves bronchospasm.

supplied by both the sympathetic and the parasympathetic nervous systems. The sympathetic nervous system is dilator to these bronchi, whereas the parasympathetic is constrictor to the bronchi. An asthmatic thus has too much parasympathetic activity, and not enough sympathetic activity.

To increase the size of the air passages, drugs such as **atropine** and **hyoscine** are given to block the parasympathetic activity, whilst drugs like **adrenaline** and **noradrenaline** are given to increase the sympathetic activity. **Isoprenaline** which is a derivative of noradrenaline is often given in the form of a spray inhalent [FIG. 86]. Since histamine also causes contraction of this smooth muscle in the bronchi and bronchioles, **antihistamine** drugs and corticoids may also be given.

7. Consolidation

In lobar pneumonia the lung may become solidified as part of the pathological changes.

PHYSIOLOGICAL PRINCIPLES UNDERLYING CLINICAL INVESTIGATIONS

Percussion. In order to ascertain the pathological changes that have occurred in these conditions, certain basic principles are used. The first is that healthy lung tissue gives a characteristic note when the chest is percussed. The note when the chest wall over healthy lung tissue is percussed is not so hollow as that found over a bubble of air in the stomach, nor so dull as that over the liver. Intermediate percussion is the method usually employed. The finger of one hand is placed firmly in contact with the chest wall and the centre of the middle phalanx struck sharply with the tip of the finger of the other hand.

This procedure may be practised using a hollow surface such as a table and a solid surface such as a wall. The difference between solid and hollow objects can be readily heard. If the right side of the thorax [FIG. 87] is percussed with the subject supine, an obvious change in note from the normal lung sound to the dull sound of the liver is heard at about the 4th right interspace. This gives the position of the diaphragm which is often higher than expected. The percussion note on the two sides of the chest at various levels front and back are compared.

Should the condition of the lung alter, there is an alteration in the percussion note. A pneumothorax gives a hollow sound whilst a pleural effusion gives a much duller sound.

Breath Sounds. The lung acts as an attenuator to sound. As a result, sounds are not readily transmitted from the large air passages through to the chest wall. When breathing is occurring, the air moving in and out of the larynx and trachea produces turbulence during inspiration, and turbulence during expiration. There are thus two breath sounds, one during inspiration and one during expiration. They can be heard clearly when a stethoscope is placed over the trachea or large bronchus. These sounds are termed **bronchial breathing.**

At the periphery of the lungs, such as in the axilla, the bronchial breath sounds are not heard. Instead a much fainter sound is heard as the air enters the alveoli. This is termed **vesicular breathing,** and consists of a faint rustling sound during inspiration and early expiration. There is no sound in late expiration. Thus bronchial breathing has two components per respiratory cycle whereas vesicular breathing has only one.

Should the lung become solidified due to collapse, or consolidated as in pneumonia, then the sound will be transmitted readily from the air passages to the chest wall and bronchial breathing will be heard over that part of the lung.

Since the sounds of air moving up and down the bronchi are not very loud, additional sounds may be produced by the subject saying '99' or '1, 1, 1'. These sounds will be modified and attenuated, so that, under normal conditions, very indistinct muffled sounds are heard when a stethoscope is applied to the chest wall. The sounds will be heard clearly if the lung is solidified.

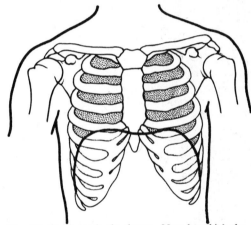

FIG. 87. The lungs in the thorax. Note how high the diaphragm extends at the front of the chest.

Low frequencies can often be felt rather than heard, so that the hand is placed flat on the chest whilst the '99' is being said, and the vibrations compared on the two sides. With consolidation there is a marked increase in the **tactile vocal fremitus,** which is the name given to these vibrations felt by the hand.

Additional sounds may also be heard. **Rhonchi** result from an excessive secretion of mucus and the large airways have a musical quality. **Râles** and crepitations arise in the terminal bronchioles and resemble a crackling noise.

Partial obstructions to breathing which are noisy may be summarized as follows:

1. Tongue obstructing pharynx = SNORING
2. Laryngeal spasm = CROWING
3. Fluid in large airways = GURGLING
4. Bronchi obstructed = WHEEZING

Clinical Examination

With these principles in mind, a patient with a respiratory disorder is investigated in the following manner. Firstly a detailed case history is taken together with full details of family and social history. Clinical examination takes the form of (1) inspection, (2) palpation, (3) percussion, (4) auscultation.

During inspection it is remembered that the two sides of the body have independent nerve supplies, and that it is very unusual for any pathological lesion of the respiratory system to affect both sides of the body equally. So that symmetry is very important and any lack of symmetry is noted. At the same time the respiratory rate and the respiratory muscles employed are seen.

The symmetrical movement of the chest can be confirmed by palpation. In order to ascertain whether the mediastinum is displaced, the trachea is palpated in the jugular notch, and the position of the apex beat located. With a pneumothorax on the right side the apex beat may be displaced to the left axilla.

Percussion and auscultation may then be used to establish the diagnosis. When percussing the back, it is important to move the scapulae round to the side and this may be achieved by asking the patient to fold the arms across the front of the chest.

Since there is a wide variation of the normal, corresponding points on the two sides of the chest are compared, since a difference between the two sides is often important diagnostically.

If the above clinical examination does not establish the diagnosis, a chest X-ray is taken and, if necessary, tests of pulmonary function and blood chemistry are carried out. These tests include the measurement of lung volumes [p. 47], vital capacity, FEV_1, functional residual capacity, maximum ventilation volume, the tensions of oxygen and carbon dioxide in the arterial blood, and the plasma bicarbonate level.

The ability of the respiratory system to move air rapidly in and out of the chest will be impaired if the lungs and chest wall lose their elasticity, if the airways are narrowed or if the respiratory muscles are impaired.

Pulmonary Compliance—'Compliance of the Lungs'

Breathing is only possible because the lungs and chest wall are elastic structures. **Compliance** is the *rather confusing* word to express their *stretchiness*. It is the measure of the *ease* with which the lungs and chest wall expand when the pressure inside them is increased.

$$\text{Thus compliance} = \frac{\text{Increase in volume}}{\text{Increase in pressure}}$$

It is the reciprocal of **elasticity** which is the increase in pressure produced by an increase in volume. Elasticity is thus the *stiffness* that exists when making the lungs and chest wall expand. Thus if the **lungs** of a patient **expand less easily,** they will have **decreased compliance** (but an increased stiffness or elasticity, hence the confusion!).

Compliance is measured by relaxing the respiratory muscles and finding the increase in tracheal pressure as the lung volume is increased by introducing air into the lungs in 500 ml. steps. This will give the combined compliance of the lungs and chest wall. It is usually measured on anaesthetized patients.

If the oesophageal pressure is also recorded as an indication of the pressure outside the lungs (but inside the chest wall = intrapleural or intrathoracic pressure) the compliance of the lungs alone can be found. From the combined compliance and the lung compliance the chest wall compliance can be calculated.

When determining compliance, volumes are measured in litres. Pressures are usually measured in cm. H_2O (1 cm. $H_2O = 0.7$ mm. Hg).

Typical figures are:

Pulmonary compliance (lungs + chest wall) = 0·1 litres/cm. H_2O.
Lung compliance = 0·2 litres/cm. H_2O.
Chest wall compliance = 0·2 litres/cm. H_2O.

The combined compliance is less than the two constituent compliances because the lungs and chest wall together are not so compliant as either the lungs or chest wall separately.

An **increased compliance** is associated with a flaccid paralysis of the muscles of the chest wall, and with an open pneumothorax.

A **decreased compliance** is due to an increased stiffness in either the lungs or chest wall. It is seen in older patients who have less elastic and more fibrous tissue, and in pregnancy due to the abdominal distension. A decrease in **chest wall compliance** will be seen when the tone in the intercostal muscles has increased with spastic paralysis [p. 132] or when pain in the ribs is inhibiting movement. It will also be seen in spondylitis of the rib joints and myositis and fibrosis of the muscles.

A decrease in **lung compliance** is seen in fibrosis of the lung, in lung oedema and in asthma.

The expansion of the lungs and chest wall during inspiration is opposed by the elastic property of the lungs, which is tending to keep the lungs as small as possible. Energy has to be expended by the respiratory muscles to stretch the lungs during inspiration, but much of this energy is recovered during the next expiration when the elasticity returns the lungs and chest wall to the resting respiratory level. Owing to friction a small amount of energy is lost as heat.

In archery, energy is expended when pulling back the bow, but this energy is recovered when the bow is released (it is transferred to the arrow which shoots off at a high velocity).

Surface Tension and Lung Compliance

The surface of water at a water-air interface behaves like a 'skin'. This property is termed **surface tension.**

This 'skin effect' allows a powder such as *flowers of sulphur* to float on the surface of water. Detergents (and soaps) lower the surface tension of watery fluids and cause the powder to sink. This fact is made use of in the *Hay's urine test* for the presence of bile salts (which cause the sulphur to sink due to their detergent-like property).

The surface tension of water impedes the passage of air through small wet tubes and air spaces. Thus it is more difficult to breathe through a gauze mask that has become saturated with water because films of water block the small air spaces in the cloth.

The **alveoli** of the lungs are *lined* with water which allows gaseous exchange to occur. (Oxygen must first dissolve in this water before it can diffuse through the alveolar wall to the blood in the pulmonary capillaries.) If during expiration so much air is expired from an alveolus that opposing sides touch, a film of water will be formed which will tend to block the entry of air at the next inspiration. An additional inspiratory effort will then be needed to overcome the surface tension and re-open the alveolus.

Surfactant. This is a detergent-like substance present in the alveoli which reduces the surface tension of the alveolar fluid. Its action is to allow the alveoli to expand more readily during inspiration and it thus has the effect of increasing the compliance of the lungs.

A **lack of surfactant** in a new-born baby (*hyaline membrane disease*) is associated with progressive pulmonary collapse, or a failure of the lungs to expand with air at birth (**atelectasis**). A surfactant deficiency may lead to areas of lung collapse following open-chest surgery in which pump-perfusion has been employed.

Resistance of the Airways to Air-flow

The air passages offer a resistance to air-flow (cf. arterioles offer a resistance to blood flow). This resistance is relatively unimportant in a normal person when the rate of air-flow is low, but it becomes important when the rate of air-flow is high in exercise or when the air passages are narrowed (bronchospasm) as in asthma.

Consider the case of a patient with bronchospasm trying to breathe rapidly. Inspiration will lower the pressure in the lungs and air will be drawn slowly into the lungs through the narrowed air passages. The first air to reach the alveoli will be alveolar air which has been left in the dead space after the previous expiration. Owing to the airway resistance it will take an appreciable time before air from the outside (room air) reaches the alveoli. With a rapid respiratory rate, expiration will have started before then, so that no room air arrives at the alveoli and there will then be no effective alveolar ventilation. This will lead rapidly to asphyxia.

Ventilation-perfusion Ratio

For an effective exchange of gases in the lungs, the alveolar ventilation must be distributed in such a way that it matches the pulmonary blood flow in all parts of the lung. Thus each part must have the correct ventilation/perfusion ratio.

To take an extreme case, if all the pulmonary blood flowed to the right lung and all the inspired air passed to the left lung, there would be no effective exchange of gases at all. The right lung blood would be 'shunted'.

Even if **shunting** occurs on a small scale there may be a marked drop in the arterial blood oxygen tension. This is because the blood flowing through the unaerated lobe of the lung will not take up oxygen and the reduced blood from this lobe will mix with oxygenated blood from the rest of the lungs. The blood leaving the lungs and passing through the left side of the heart to the arteries will contain reduced haemoglobin. If more than 5 grams of haemoglobin in the blood are in the reduced form the patient will be visibly cyanosed [p. 57].

Owing to the slope of the oxygen dissociation curve and the fact that the blood leaving the normal lung area is already very nearly completely saturated with oxygen, it is not possible for the normal aerated lung area to take up more oxygen to compensate. Nor will giving the patient an enriched oxygen mixture to breathe be effective. It will only increase the oxygen in solution and have little effect on the quantity of oxygen carried as oxyhaemoglobin.

There is usually no carbon dioxide retention because the carbon dioxide dissociation curve has a different shape and more is excreted by the aerated lung tissue.

A mechanism exists whereby the blood flow to a lobe of the lung is cut off if the lobe ceases to be aerated. The way in which this is brought about is obscure, but it is probably a local response of the pulmonary capillaries which constrict when there is a lack of oxygen in the alveoli.

Summary

A **high** P_{CO_2} in the arterial blood is an indication of **insufficient alveolar ventilation.** This may be due to depression of the respiratory centre, respiratory muscle weakness, lung and chest wall lesions, an obstruction to breathing or an increase in the physiological dead space due to under-perfused alveoli.

Shunting is suspected when the P_{O_2} in the **arterial blood** is markedly **lower** than the P_{O_2} in the **alveolar air** whilst the P_{CO_2} is normal.

EMERGENCY RESUSCITATION

Should a person stop breathing, then it is essential to start **artificial respiration** as quickly as possible. The most effective 'first-aid' artificial respiration method available is the mouth-to-mouth expired air resuscitation technique. In order to carry out this method satisfactorily, it is important to establish a clear airway, since an unconscious person may very easily swallow his tongue which will block the pharynx [FIG. 88]. The subject is placed on his back, and the head should be extended, the nose closed, and the jaw moved forwards as in FIGURE 89. The mouth is firmly applied to the subject's mouth, and the lungs inflated with the operator's own expired air. The expired air still con-

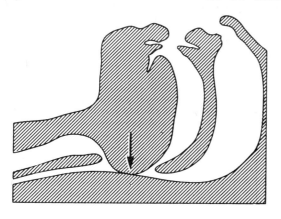

TONGUE BLOCKING PHARYNX
Air passages shown in white

FIG. 88. When an unconscious person is lying on his back the tongue may flop backwards and block the airway.

tains 16–17 per cent. oxygen (instead of room air, 21 per cent.) and this amount of oxygen is quite adequate for the respiration. After inflating the subject's lungs the operator should lift his head away, and turn so as to watch the subject's chest. If a clear airway has been established, the chest will be seen to deflate. At the same time another inspiration is taken and the mouth applied to the subject's mouth for the next inflation.

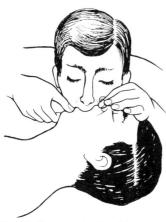

FIG. 89. Mouth-to-mouth breathing (expired air resuscitation). The nose is closed by the thumb and forefinger and the hand pressed on the forehead to extend the head. In addition the lower jaw is pulled forward. The lungs may now be inflated by applying the mouth to the subject's mouth. Alternatively, the lungs may be inflated through the nose.

If six inflations of the lungs do not produce any improvement in the colour of the subject, then a **cardiac arrest** should be suspected. It is important to establish whether the heart is beating or not, before cardiac re-

suscitation is started. The signs to look for, are the absence of the carotid pulse, an absence of heart sounds (as heard by applying one's ear to the subject's chest), and dilated pupils (although this may be found in other conditions).

External cardiac massage consists of applying a pressure, using the heel of one hand with the other hand on top [FIG. 90], to a point in the midline one third the way

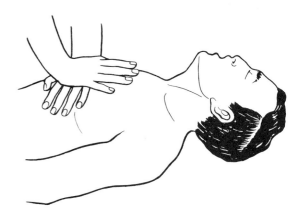

FIG. 90. External cardiac massage. Pressure is applied with the heel of the hand only to the sternum in the midline at the level of the 4th interspace (in line with the nipples in the male).

up the sternum. The sternum is depressed by a distance of $1-1\frac{1}{2}$ inches, and, by so doing, blood is forced from the underlying ventricles out into the arteries. On relieving this pressure, the blood enters the ventricles from the atria and veins, and by repeating this procedure 60 or so times a minute, an adequate circulation can be maintained. Unfortunately it is not possible to maintain an adequate pulmonary ventilation of the lungs by this technique, and it is essential to carry out **mouth-to-mouth resuscitation** alternating with the **external cardiac massage.** One lung inflation is given, followed by 6 cardiac presses, followed by another lung inflation and so on. Carried out in this manner it is possible to maintain the life of a patient for one hour or longer.

Cardiac massage must be started within 2 minutes of a cardiac arrest. Otherwise there may be irreversible brain damage.

Subsequent treatment will depend on the services available. Artificial respiration can be maintained using a pump, following an **intubation** of the air passages through the mouth, or following a **tracheostomy.** If the heart is in a state of **ventricular fibrillation,** defibrillator electrodes can be applied across the chest to send an electric shock to the heart so as to revert the heart to normal rhythm.

An **external cardiac pacemaker** can be used to stimulate ventricular contraction after passing an electrode into the right ventricle via the veins of the neck [see p. 10].

It will probably be necessary to **correct the acidaemia** of the blood (that is the blood pH may have fallen from 7·4 to 7·2 or lower as a result of the formation of lactic acid and other metabolites due to the oxygen lack). This correction is usually carried out by setting up an intravenous infusion whilst the cardiac massage is in progress and running in 150 milli-equivalents of sodium bicarbonate for every minute of cardiac arrest prior to starting the cardiac resuscitation.

When the patient recovers he should be turned to the semi-prone coma position with the head to one side to prevent the inhalation of any vomit [FIG. 91].

FIG. 91. Coma position. When the subject starts to recover he should be turned to the coma position to prevent inhalation of vomit.

Coma and Respiratory Obstruction

An oropharyngeal airway is used in light coma to prevent the tongue obstructing the respiration. With deep coma where reflexes are absent or when a respiration is depressed, it is usual to intubate using a cuffed endotracheal tube. This allows positive pressure artificial respiration to be employed if necessary. Such intubation tubes are not normally left in position for longer than 48 hours.

The technique of tracheal intubation for inhalation anaesthesia and positive pressure mechanical ventilation has enabled muscle relaxant drugs to be used safely and has made open heart surgery possible.

11. METABOLISM

Man needs food in order to stay alive. The complex process whereby this food is converted into heat and energy and used for the growth and repair of tissues is termed **metabolism.** But before the food can be used by the cells, it has to be transformed into a form suitable for absorption into the blood. This conversion is termed **digestion.** Only the carbohydrates, fat and proteins in the diet need to be changed in this way. The inorganic salts, water and the vitamins are absorbed without undergoing any chemical change.

Enzymes

Digestion is brought about by biological catalysts known as **enzymes.** These are themselves proteins. Using the modern terminology, the name of an enzyme is derived from the name of the substance on which it acts, but with the ending of the word changed to the letters **-ase.** Thus the enzyme in the small intestine which acts on **sucrose** (cane or beet sugar) is termed **sucrase.** That acting on **lactose** is termed **lactase.** Occasionally the Latin form of the word is used. The Latin for starch is *amylum*, and the enzyme which breaks down starch is termed **amylase.** Protein chains are broken down by proteases into shorter peptide chains. **Peptides** are broken down further and finally to amino acids by **peptidases.**

The older names of some digestive enzymes are so well established that they are still in common use. These include **ptyalin** (salivary amylase), **pepsin** (gastric protease), **trypsin** and **chymotrypsin** (pancreatic peptidases).

The distinction between **peptides** and **proteins** is vague. One criterion is to consider the number of amino acids in the chain. If it is over 100, the substance is a protein. If less than 100 amino acids, it is a peptide. But shorter chains may occasionally be referred to as proteins, for example, insulin (51 amino acids).

PROTEIN

Protein is needed in the diet for growth and repair of tissues. Body protein is made up of long chains containing many hundreds of amino acids. Only twenty different amino acids are used by the body in the formation of the protein molecule, so each of the 20 different amino acids will be used many times.

The twenty amino acids are:

Alanine	*Leucine
Arginine	*Lysine
Asparagine	*Methionine
Aspartic acid	*Phenylalanine
Cysteine	Proline
Glutamic acid	Serine
Glutamine	*Threonine
Glycine	*Tryptophan
Histidine	Tyrosine
*Isoleucine	*Valine

Although 20 different amino acids are employed in the formation of body protein, only 8 of these are essential amino acids and must be present in the food protein. These are marked with an asterisk. Provided that these are present in the diet, the remaining twelve can be made by the liver.

Although these compounds are termed *acids*, the majority of the **amino acids** are **neutral** substances which will make the blood neither more acid nor more alkaline. The exceptions are: aspartic acid and glutamic acid which are acidic, and arginine, lysine and histidine which are basic (alkaline). All the others contain one acidic carboxyl group ($-COOH$) and one basic amino group ($-NH_2$). These groups tend to neutralize one another.

Methionine and **cysteine** are amino acids which contain **sulphur.** As will be seen later, the sulphur in any surplus methionine and cysteine is converted in the body to sulphuric acid. This sulphuric acid is excreted by the kidneys and appears in the urine (in buffered form). If the kidneys are unable to excrete this sulphuric acid, as in kidney failure, acidaemia will develop unless the protein intake is drastically reduced [p. 86].

Digestion of Food Protein

The process whereby the amino acids are obtained from food protein is termed protein digestion. We could eat the 20 amino acids each day as chemicals, but instead we find some animal or plant that has used the same amino acids in forming its own protein and we eat the meat of the animal or part of the plant.

The digestion of protein starts in the **stomach** where the enzyme **pepsin,** in the presence of **hydrochloric acid,** starts the breakdown of the food protein into shorter-chained molecules known as peptides. The food then passes into the duodenum where it comes into contact with the **pancreatic juice** which contains the enzymes **trypsin** and **chymotrypsin.** These enzymes break the chains still further:

FOOD PROTEIN	$\xrightarrow[\textbf{stomach}]{\textit{pepsin}}$	shorter chains (peptides)	STOMACH
	$\xrightarrow[\substack{\textit{chymotrypsin} \\ \textbf{duodenum}}]{\textit{trypsin}}$	very short chains	DUODENUM
	$\xrightarrow[\textbf{small intestine}]{\textit{peptidases}}$	amino acids	SMALL INTESTINE

The digestion is completed in the **small intestine** where a **collection of peptidases** known as *erepsin* completes the breakdown to amino acids. In the small intestine, some peptidases remove the amino acids from one end of the peptide chain, whilst other peptidases act on the centre of the chain or at the other end. One end of the peptide chain has a free amino group (—NH₂) whilst the other end has a free carboxyl group (—COOH). The peptidases which act at the amino end are termed **aminopeptidases,** whereas the peptidases that act at the other end are termed **carboxypeptidases.** Erepsin is a collection of these and similar peptidases.

Amino Acids for Growth and Repair of Tissues

By the time the pepsin in the stomach, the trypsin and chymotrypsin from the pancreatic juice and the peptidases of the small intestine have acted, the food protein has been broken down into the individual amino acids which are then available for absorption.

The juicy steak, the egg or the milk protein have all changed into the same 20 amino acids. These amino acids are absorbed into the blood in the small intestine and they circulate round in the blood. Thus a drop of blood from a finger prick will contain in its plasma the 20 amino acids that have been absorbed from the previous meal.

These amino acids circulate in the blood and every cell in the body extracts from the blood the amino acids it needs for growth and repair [FIG. 92].

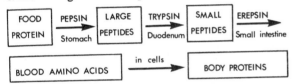

FIG. 92. Food protein is broken down by the process of digestion to amino acids which circulate in the blood and are used by the cells to build body proteins.

If the skin has been cut, then the skin cells will make new cells to bridge the gap. The cells on either side of the lesion take amino acids from the blood and join them in the correct order to form skin protein. At the same time, cells in the bone marrow, for example, are removing the same amino acids and joining them together to form blood cells. Growth of the body is based on the same amino acids which come from the ingested food protein.

When planning a diet it is essential to ensure that all the eight essential amino acids are present in the dietary protein. Otherwise growth and repair of tissues will be impaired.

All protein from animal sources (meat, fish, milk, etc.) contains the eight essential amino acids. But protein from vegetable sources may be relatively deficient in one or more of the essential amino acids. For example, wheat flour is relatively deficient in lysine, whereas the protein in maize is also relatively deficient in tryptophan and tyrosine. Thus bread and cereal foods derived from maize may have a deficiency of essential amino acids. Milk, on the other hand, contains all the essential amino acids, and if milk is taken at the same time, then all the amino acids will be supplied.

It is possible to supply all the essential amino acids from vegetable sources. In fact, millions of people live entirely on a vegetarian diet. Rice, for example, contains all the essential amino acids, and it is possible to live on rice and water. The growing part of a plant usually contains the eight essential amino acids. There is thus a simple rule to ensure that a vegetarian diet contains the essential amino acids, and that is to make certain that the food eaten contains that part of a plant by means of which it reproduces itself. Peas will grow if planted, and so will beans, nuts and lentils, and these are all first class proteins which will supply the eight essential amino acids. Soya bean meal forms a valuable supplement to a rice diet.

Cellulose cannot be Digested

The process of digestion in man is limited by the fact that he cannot digest **cellulose.** Cellulose is a very common substance in the vegetable kingdom since it forms the structure of plants. As a result much of the protein and carbohydrate made by the action of sunlight on plants is not available directly to man.

The paper on which this book is printed is a form of cellulose. Chemically it has as much glucose in it as starch has, but paper cannot be digested because there is no enzyme cellulase in the digestive tract.

The **potato** consists of **starch grains** surrounded by a **cellulose envelope.** If a raw potato is eaten, even if it is chewed thoroughly, it will pass through the digestive tract and appear undigested in the faeces. A raw potato therefore has no food value. On the other hand if the potato is heated, then the cellulose envelope will burst, and the content is then available for digestion. This is an example of a food which has no food value in the raw state, but has a very high food value in the cooked state.

Sources of Protein

With the increase in the world population, a major problem is arising concerning the feeding of so many people. The amino acids are derived from the action of sunlight on plants and thus all protein has a vegetable origin. Because of the absence of cellulase in the digestive tract of man, most of this protein in plants is not directly available.

Instead of eating grass, which would pass through the digestive tract undigested, we allow an animal to eat the grass and then we eat the animal or drink the animal's milk. A cow has cellulase in its digestive tract, and is able to extract the amino acids and carbohydrate from grass. The cow's milk contains protein derived from the grass amino acids.

The sunlight on the surface of the sea causes the growth of plankton, which is eaten by small fish. These are eaten by larger fish and finally we eat the larger fish as protein.

In the not too distant future, it will be essential to find a way of by-passing the animal intermediary and to enable protein and carbohydrate to be extracted directly from the vegetable sources.

In an attempt to solve the world's shortage of protein, textured foods are being developed using thermoplastic die extrusion of concentrates of vegetable proteins such as the soya bean. Protein filaments can be made which mimic the texture of meat and after the addition of flavouring, etc. may prove to be an acceptable substitute. The protein content will be as high as, if not higher, than the meat that they are replacing.

Abnormal Haemoglobin due to One Amino Acid Error

The arrangement of amino acids in body protein is usually unique to a given person. Thus the skin of one person is slightly different from the skin of another, and a skin transplant will be rejected.

Some defects in protein formation are inherited. If the amino acid *valine* is incorporated in the haemoglobin molecule in place of *glutamic acid* at the sixth amino acid position of one of the protein chains, then although the remaining 139 amino acids are correct, sickle-cell anaemia will develop. In this condition the red cells take up an irregular shape when the haemoglobin is in the reduced form (not carrying oxygen). The cells are then haemolysed leading to anaemia.

SURPLUS PROTEIN

In a well-developed country more protein is eaten than is required for growth and repair of tissues. The surplus protein is used for **heat and energy.**

As has been seen [p. 45] heat and energy are obtained in two ways:

1. Carbon in the food is oxidized with oxygen from the air to form heat and energy and the waste product carbon dioxide.

2. Hydrogen in the food is oxidized with oxygen from the air to give heat and energy and the waste product water.

The surplus protein is converted into carbon dioxide, water, heat and energy, but the amino acids contain another element as well as carbon, hydrogen and oxygen. This is the element nitrogen. The nitrogen has to be removed and excreted as a waste product. In man, the nitrogen is converted to urea and this urea passes out via the kidneys dissolved in water as urine [FIG. 93].

About **30 grams of urea** are made each day from the **surplus protein** on a normal diet. On a high protein diet the amount may be greatly in excess of this. In such a situation a large volume of water would be needed to excrete the large amount of urea produced. The volume of urine produced would therefore increase markedly.

Removal of the nitrogen part of the amino acid molecule is termed **deamination.** The nitrogen is converted first to ammonia[3], and in the **liver** the ammonia is converted to urea.

[3] On standing, urine may develop an ammoniacal smell due to the conversion of the urea back to ammonia by the enzyme *urease* present in bacteria.

In liver failure the ammonia is still formed, but it is not converted to urea. As a result ammonia intoxication may occur, leading to coma.

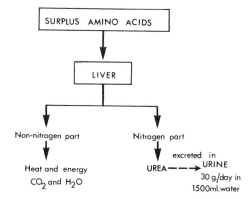

FIG. 93. The amino acids which are surplus to the body's requirements for growth and repair of tissues are deaminated. The nitrogen part is converted to urea which is excreted in the urine. The carbon and hydrogen part is used for heat and energy.

Krebs Urea Cycle

Although deamination and the formation of ammonia takes place in all the cells of the body, conversion of ammonia to urea occurs only in the liver. It is formed by a sequence of events which is known as the Krebs urea cycle [FIG. 94]. The over-all reactions are:

$$CO_2 + 2NH_3 \rightarrow CO(NH_2)_2 + H_2O$$

$$\underset{\text{carbon dioxide}}{} \quad \underset{\text{ammonia}}{} \quad \underset{\text{urea}}{} \quad \underset{\text{water}}{}$$

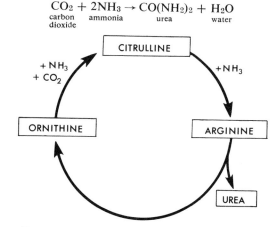

FIG. 94. Krebs urea cycle (simplified). The formation of urea takes place only in the liver.

During growth the nitrogen intake exceeds the nitrogen loss (since it is incorporated in the body proteins), but once growth has been completed, the amount of nitrogen gained from the food equals the nitrogen lost in the urine after allowing for the loss in desquamated skin, shed intestinal cells in the faeces, hair, finger nails, etc.

Anyone who is excreting more nitrogen in the urine than he is gaining from the food amino acids, is in 'negative nitrogen balance'.

One gram of protein used for heat and energy gives 4 Calories of heat and energy. If the food protein is used for growth and repair of tissues it will, of course, provide *no* Calories at all.

CARBOHYDRATES

The term carbohydrate is used to describe the starches and sugars which are ingested to supply the body with heat and energy. They are compounds containing carbon, hydrogen, and oxygen only.

One gram of carbohydrate provides 4 Calories of heat and energy.

Digestion of Carbohydrate

Starch (amylum) is broken down to **maltose** and then to **glucose** by the process of digestion. The enzyme **amylase,** which breaks down starch to *maltose,* is found in saliva and in the pancreatic juice. The salivary amylase acts for a comparatively short time because the food is swallowed and on entering the stomach, the hydrochloric acid penetrates the bolus and makes it too acid for the salivary amylase to continue its action. Starch digestion then ceases, and does not start again until the food has reached the duodenum. Here the pancreatic juice which is entering the duodenum via the pancreatic duct completes the breakdown to maltose. The breakdown of maltose to glucose is brought about by the enzyme maltase which is found in the pancreatic juice and also in the small intestine. The breakdown of maltose to glucose in the small intestine, by the action of **maltase,** probably occurs inside the cells which form the villi of the small intestine.

Cane or beet sugar (sucrose) is broken down in the small intestine by the enzyme **sucrase** into *glucose* and *fructose.* All the cells of the body are able to utilize fructose and to convert it into glucose.

The sugar in milk (lactose) is broken down by the enzyme **lactase** in the small intestine to *glucose* and *galactose.* Galactose is converted to glucose by the cells of the liver.

Only the liver is able to carry out this conversion of galactose to glucose and this fact is used as a **liver function test.** Galactose is given and the time taken for it to disappear from the blood is noted. In a normal person all the galactose disappears in 15 minutes, but in cases of liver disease galactose may still be present in the blood after one hour.

Some babies are born without the necessary enzyme in the liver to convert galactose to glucose. The galactose then accumulates in the blood, since it cannot be used by the other cells in the body, and this gives rise to the condition of **galactosaemia.** This is an *inborn error of metabolism* which is inherited from the parents as a recessive, that is, the parents themselves may not have shown any outward signs of suffering from this condition. The toxic effects of galactosaemia include failure of brain development leading to mental retardation. The galactose appears in the urine where it may be detected. A baby with this condition, at least in the early stages of life, must not be given any milk and must be fed on a lactose-free diet.

$$STARCH \xrightarrow{amylase} MALTOSE \xrightarrow{maltose} GLUCOSE$$

$$SUCROSE \xrightarrow{sucrase} GLUCOSE + FRUCTOSE$$

$$LACTOSE \xrightarrow{lactose} GLUCOSE + GALACTOSE$$

Phenylketonuria and Maple Syrup Urine Disease

Phenylketonuria and maple syrup disease are two further inborn errors of metabolism found in babies which will give rise to mental changes if untreated. The inherited defect in this condition is the body's inability to deal not with surplus galactose but with surplus amino acids.

In **phenylketonuria,** the amino acid phenylalanine accumulates in the blood producing toxic effects. The kidneys excrete the phenylalanine as a derivative (a phenyl-ketone) and it is this substance that may be detected in the urine.

In **maple syrup urine disease,** it is the amino acids leucine, isoleucine and valine, that accumulate in the body and produce severe mental changes. The derivatives of these amino acids in the urine have a smell similar to that of maple syrup, hence the name of the disease.

In both these conditions, the protein intake must be restricted, so that there will not be a surplus of the amino acids involved. Alternatively, synthetic diets containing known amounts of each amino acid may be employed.

Lactose Intolerance

In many races, who do not normally ingest milk as an item of diet, the level of **lactase** in the small intestine declines after $1\frac{1}{2}$ years of life, and is virtually absent in the adult. When milk is given to such a person, the lactose remains undigested and unabsorbed in the gut. It causes the retention of water in the intestines by its osmotic action and this, coupled with the production of large volumes of carbon dioxide by bacterial action, leads to intestinal upsets.

METABOLIC PROCESSES INVOLVED IN THE PRODUCTION OF HEAT AND ENERGY

It has been seen that the process of digestion of food carbohydrates results in the entry into the blood of glucose. The blood glucose level is maintained at 60–100 milligrams of glucose per 100 ml. of blood. This level is very important for the functioning of the brain, since the brain cells use glucose as their food.

Should the blood glucose level fall to 40 milligrams of glucose per 100 ml. blood, the brain will stop functioning and the person will go into a hypoglycaemic coma.

When an excess of glucose enters the blood from the digestive tract the surplus is converted to liver glycogen [FIG. 95]. This conversion is aided by the release of the hormone **insulin** into the blood from the pancreas. When required to maintain the blood glucose level, the liver

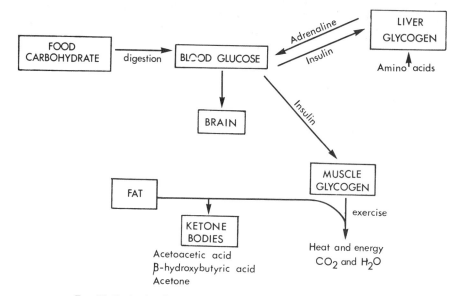

FIG. 95. Basic plan for the utilization of carbohydrate and fat.

glycogen will be converted back to glucose. This conversion back is facilitated by the hormones **adrenaline** from the adrenal medulla, and **glucagon** from the pancreas.

Liver glycogen can also be made from non-carbohydrate sources, that is, from the deaminated amino acids, and from fat. This process is termed **neoglucogenesis.**

If so much carbohydrate is eaten that the liver glycogen store becomes full, then the surplus blood glucose is converted to fat. This fat is deposited in the fat depots of the body, such as in the adipose tissue under the skin, and at the back of the abdomen.

Muscle Glycogen

Blood glucose is converted, with the aid of insulin, to muscle glycogen. Muscle glycogen provides the source of heat and energy for muscular activity. When required it is broken down to carbon dioxide and water, and heat and energy.

Muscle glycogen, unlike liver glycogen, is not available to maintain the blood glucose level.

The breakdown of muscle glycogen to form heat and energy takes place in two stages. In the first stage the muscle glycogen is broken down to pyruvic acid (which contains three carbon atoms). If an adequate oxygen supply is available, the pyruvic acid is broken down in the second stage to carbon dioxide and water. If no oxygen is available the pyruvic acid is converted instead to lactic acid. This enables more pyruvic acid to be formed. The production of heat and energy with the formation of lactic acid, which occurs in the absence of oxygen, is called **anaerobic metabolism.** The body is then said to have developed an 'oxygen debt'. This oxygen debt has to be repaid after the exercise has been comple-

ted. Thus it is possible to run a short distance (100 yards) without breathing. The energy comes from the formation of lactic acid. After the exercise has been completed the subject breathes deeply and rapidly for the next few minutes in order to take in oxygen to 'repay' the oxygen debt. About one fifth of the lactic acid formed is converted to carbon dioxide and water when the oxygen is available, and the remaining four-fifths are converted back to muscle glycogen for re-use on another occasion.

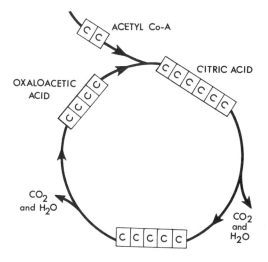

FIG. 96. Krebs citric acid cycle (simplified). The conversion of acetyl co-A to carbon dioxide and water with the release of heat and energy. Each rectangle containing the letter C represents a carbon atom.

Aerobic Metabolism

If oxygen is available then the pyruvic acid is converted first to an active form of acetic acid (acetyl co-A), which contains two carbon atoms. This acetyl co-A enters the Krebs citric acid cycle [FIG. 96]. A four-carbon acid (oxaloacetic acid) joins up with the two-carbon acid (acetyl co-A) forming a six-carbon acid (citric acid). This then becomes a five-carbon acid, with the formation of carbon dioxide and water, which then becomes oxaloacetic acid again (four-carbon acid) with the formation of carbon dioxide and water. Oxaloacetic acid is then available again to oxidize more acetyl co-A. Such a sequence of events is termed a cycle.

Adenosine Triphosphate (ATP). The energy produced by the oxidation of glycogen to carbon dioxide and water is used first to form the energy-rich compound ATP (adenosine triphosphate). When required for muscular contraction, and for chemical reactions such as the synthesis of proteins, the energy stored in ATP is released as it changes to ADP (adenosine diphosphate).

FATS

Fats are compounds containing carbon and hydrogen with very little oxygen. They are a very concentrated form of food. **One gram of fat gives 9 Calories of heat and energy.**

Fats are compounds of glycerol (glycerine) and three fatty acids [FIG. 97]. They are triglycerides.

Fats are not water soluble and fat digestion is the process whereby the physical state of the fat is changed to enable it to mix with water. In the duodenum and small intestine the fat is emulsified by the combined action of the bile salts and the pancreatic lipase. The bile salts are the body's detergent and enable the fat to mix with water. The lipase tends to split the fat into glycerol and fatty acids, but only a small amount of the fat ingested is completely split in this way. Some of it

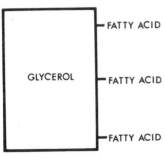

FAT (TRIGLYCERIDE)

FIG. 97. Neutral fat is a compound of glycerol and three fatty acids. It is a triglyceride.

has one or two amino acids removed leaving a di-glyceride or a monoglyceride [FIG. 98] which is more soluble in water than the original fat. These glycerides are absorbed into the blood stream and pass via the portal vein to the liver.

The majority of the emulsified fat is absorbed, not into the blood, but into the lymphatics. It passes up the thoracic duct, and enters the great veins in the neck as fat droplets.

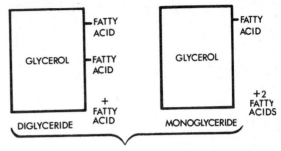

after the action of lipase

FIG. 98. The enzyme lipase splits off one or more of the fatty acids from fat. If two fatty acids remain the resultant compound is a diglyceride. If one remains the resultant compound is a monoglyceride. (If none remain the resultant compound is glycerol (glycerine).

The characteristic appearance of milk is due to the fact that it consists of fat droplets in water. The contents of the intestinal lymphatics have a similar milky appearance and for this reason the lymphatic vessels are termed **lacteals.**

Since the fat has not undergone any chemical change, the fat entering the blood stream from the thoracic duct will be the same as the fat ingested. Thus the blood after a fatty meal may contain butter fat droplets, margarine fat droplets or olive oil droplets, depending on the type of fat ingested.

The fats are converted into body fat by the liver, and are stored in the adipose cells of the body. At body temperature, these fats are liquid (and strictly speaking should be termed 'oils'). When required the fat is broken down to glycerol and fatty acids by lipase in the tissues. It is then utilized by the body cells for heat and energy.

Fat Metabolism

Provided that carbohydrates are being metabolized, the fat which is broken down to acetyl co-A will enter the Krebs cycle [FIG. 96]. Fat, however, is unable to make oxaloacetic acid, and if no carbohydrate metabolism is proceeding, then the acetyl co-A from the fat will not be used. Instead the acetic acids join up in pairs and form acetoacetic acid. Some of the aceto-acetic acid changes into beta-hydroxybutyric acid and acetone. These three substances are termed **ketone bodies.**

Ketosis

Thus if fat is metabolized without the concurrent metabolism of carbohydrate, ketone bodies accumulate in the blood and give rise to **ketosis.** These ketone bodies are toxic to the brain and, if they reach a high enough blood level, will give a **ketotic coma.**

Ketosis occurs when a large amount of fat is ingested without an appropriate amount of carbohydrate. It also tends to occur in starvation where there is a deficiency of carbohydrate in the diet, and body fat is being utilized.

Diabetes Mellitus

The most severe forms of ketosis arise in cases of **diabetes mellitus.** In this condition the pancreas does not produce enough insulin to allow sufficient glucose to enter the cells. The formation of muscle glycogen is deficient and there is a reduction in the carbohydrate metabolism. As a result much of the blood glucose will be unused and the blood glucose level will rise.

As will be seen later [p. 85] when the glucose level in the blood exceeds about 180 milligrams of glucose per 100 ml. blood, the kidney tubules are unable to re-absorb all the glucose, and glucose appears in the urine. Water is also excreted with this glucose, and the person becomes thirsty. He passes a large volume of urine and may become dehydrated. The term *diabetes* means that a large volume of urine is passed. The term *mellitus* (= sweet) dates from the time when the urine was tested by tasting and the urine in this condition is sweet to the taste [see diabetes insipidus, p. 79].

It is the disorder of fat metabolism that ultimately causes an untreated diabetic patient to go into coma. Since carbohydrate is not being metabolized, the fat will also be unused and will be converted, not to carbon dioxide and water, but to ketone bodies. As has been seen this will lead to a **ketotic (diabetic) coma.**

Glucose Tolerance Test (G.T.T.)

A patient who is suspected of having diabetes is usually given a glucose tolerance test. As its name implies, this is a test to see if a patient can tolerate the ingestion of glucose (usually 50 grams) without the blood glucose reaching too high a level or glucose appearing in the urine. Blood samples are taken when the glucose is given and at intervals over the next $2\frac{1}{4}$ hours. The blood is put into a tube containing fluoride as the anticoagulant. This not only forms insoluble calcium fluoride and thus removes the calcium ions, but it also 'poisons' the enzymes that would otherwise cause the glucose to be used up whilst the tube is waiting for analysis.

The bladder is emptied at 1 and 2 hours and the urine tested for glucose.

The **prednisone tolerance test (P.T.T.)** is a similar test, but is based on the fact that adrenal corticoids oppose the action of insulin and switch the body's metabolism from using carbohydrate to using protein [p. 104]. Since the utilization of glucose by cells falls, its blood level rises, and in patients with *latent diabetes* glucose may appear in the urine. This test lasts about 18 hours.

Treatment of Diabetes

A diabetic patient is treated by injections of insulin which are adjusted to match the carbohydrate intake. Since it would be inconvenient to have to weigh each meal, calculate the carbohydrate content and to inject the corresponding amount of insulin required, a diabetic patient is usually on a diet and has fixed injections of insulin each day.

Insulin (a 51 amino acid peptide) is given by subcutaneous injection. By mouth it would be broken down by the digestive enzymes to its constituent amino acids. Although it has been chemically synthesized, insulin for injection is extracted from the pancreas of animals. *Soluble insulin* is absorbed rapidly. This absorption can be delayed by combining the insulin with a protein such as protamine (as in *protamine zinc insulin*). A mixture of rapidly and slowly absorbed forms is often used to increase the time over which an injection is effective and thus minimize the number of injections required daily.

Should an excessive dose of insulin be taken, or alternatively should the injection of insulin be given and no carbohydrate taken to match it, then the injected insulin will lower the blood sugar level and the person may go into a **hypoglycaemic (insulin) coma.**

A diabetic patient is thus between two forms of coma, a ketotic coma if insufficient insulin is taken, and a hypoglycaemic coma if too much insulin is taken. Diabetic patients are usually provided with a means of testing their urine. Since ketosis comes on more slowly than hypoglycaemia, the insulin injections are usually adjusted to keep the blood glucose at a slightly higher level than normal, and to allow a small amount of glucose to appear in the urine.

Oral Antidiabetic Drugs

Certain sulphonamides (derivatives of sulphonylurea) have been found to lower the blood glucose level. They probably act by stimulating the pancreas to release insulin, and thus raise the blood insulin level. These compounds include *tolbutamide, chlorpropamide* and, more recently, *glybenzcyclamide.* Since these drugs are effective by mouth, they enable certain diabetic patients to be treated without the need for insulin injections.

12. NUTRITIONAL REQUIREMENTS

The diet must supply the body each day with:

1. Adequate Calories and Amino Acids
2. Vitamins
3. Mineral salts
4. Water.

CALORIES AND AMINO ACIDS

The heat and energy requirements of the body are expressed in Calories. This is spelt with a capital C and is the amount of heat required to raise the temperature of one litre of water (1 kg.) by one degree centigrade. It is equal to 1,000 calories (which is spelt with a small c) which is the amount of heat required to raise the temperature of only one gram of water one degree centigrade.

The Calorie is not commonly encountered in everyday life so it is useful to remember that one Calorie is approximately equal to one watt-hour. One thousand Calories are approximately equal to one kilowatt hour.[4]

Thus, if ten people are in a room, and each is giving off 100 Calories of heat per hour, the room will be receiving a total of one thousand Calories an hour, and this is approximately the same as the heat given off by a one kilowatt electric fire. It is for this reason that the temperature of a room will rise if it is occupied by a large number of people, and there is insufficient ventilation.

Determination of Energy Requirements

The total Calorie requirements of a man could be determined by measuring the heat he gives off in 24 hours. This method is not frequently employed because the subject has to be placed in a **respiration calorimeter**, that is, a chamber in which he can live for an extended period of time, and the heat he produces measured.

The Calorie requirements are usually measured indirectly. It has been found that every litre of oxygen taken into the body produces 5 Calories of heat and energy. It is thus only necessary to determine the oxygen uptake per minute.

The oxygen uptake per minute is determined by getting the subject to breathe oxygen from a spirometer and noting the rate at which the oxygen is used up. The carbon dioxide which is produced is absorbed by means of soda-lime.

Alternatively, the oxygen uptake per minute may be determined using a Douglas bag for the collection of expired air. Since the percentage of oxygen in the inspired room air is known, then a determination of the percentage of oxygen in the expired air and a measurement of the volume of expired air per minute will enable the oxygen uptake to be calculated.

[4] More accurately, 860 Calories equal one kilowatt hour. 1 Calorie equals 4,186 joules $\simeq$ 4·2 kilojoules.

EXAMPLE: Thus if the inspired air contains 21 per cent. oxygen and the expired air contains 16 per cent. oxygen, there has been a drop of 5 per cent. in the oxygen concentration.

If the pulmonary ventilation is 5 litres per minute, the oxygen used up will be 5 per cent. of 5 litres which equals 250 ml. per minute.

1,000 ml. oxygen used per minute produces 5 Calories of heat and energy.

Hence 250 ml. per minute corresponds to $\dfrac{5 \times 250}{1,000}$ Cals.

Thus the metabolic rate $= 1\frac{1}{4}$ Calories/minute

$= 75$ Calories/hour.

Basal Metabolic Rate

The basal metabolic rate is the Calorie requirement under basal conditions. This is a state of complete mental and physical rest 12–18 hours after a meal so that digestion and absorption have been completed.

The basal Calorie requirements depend on the size of a person, a large person needing more Calories per hour than a small person. But, it has been shown experimentally that this Calorie requirement is not directly proportional to the body weight (as would be expected), but is closely related to the surface area of the skin (skin surface area).

For this reason for comparison purposes between one person and another, basal Calorie requirements are usually expressed, not as Calories per hour, but as Calories per square metre of body surface area per hour. The body surface area is determined from the height and weight [TABLE 2].

TABLE 2

Body Surface Area in Square Metres for Different Body Heights and Weights

		WEIGHT		
		7 stone (98 lb.) (45 kg.)	10 stone (140 lb.) (64 kg.)	13 stone (182 lb.) (83 kg.)
HEIGHT	5 ft (1·52 m.)	1·4	1·6	1·8
	5 ft 6 in. (1·68 m.)	1·5	1·7	1·9
	6 ft (1·83 m.)	1·6	1·8	2·0

The basal requirements calculated in this way are 40 Calories per square metre per hour for a man and 37 Calories per square metre per hour for a woman.

The body surface area for an adult normally lies between 1·4 and 2 square metres with an average of about 1·8 square metres.[5] Thus, 40 Calories per square metre per hour for a 1·8 square metre man would work out at a total of 72 Calories per hour, or 1,700 Calories per day.

High metabolic rates are found in thyrotoxicosis whilst low metabolic rates are found in myxoedema [p. 102].

Calorie Requirements per Day

We have just seen that the basal requirements for a man of surface area 1·8 square metres are 1,700 Calories per day. This rate of metabolism is approximately equal to the Calories required during the 8 hours of sleep (560 Calories). During the 16 hours awake, further Calories will be required for exercise. An additional 200 Calories per hour are required when walking about. No additional Calories are required for brain activity!

If we allow an additional 1,000 Calories for the waking hours, and protein (amino acids) equivalent to 100 Calories for growth and repair of tissues [p. 65], this brings the total to 2,800 Calories per day.

1 gram of carbohydrate gives 4 Calories.
1 gram of fat gives 9 Calories.
1 gram of protein gives 4 Calories when it is used for heat and energy.

Thus a typical diet might consist of:

375 grams of carbohydrates = 1,500 Calories
100 grams of protein = 400 Calories
100 grams of fat = 900 Calories
Total 2,800 Calories.

It should be remembered when planning a diet that although sugar is pure carbohydrate, meat contains 70 per cent. water and hence it contains less than 30 per cent. protein. Butter contains 20 per cent. water and 80 per cent. fat.

Starvation

If insufficient food is being eaten, then the tissues of the body are used for the production of heat and energy. The glycogen store of carbohydrate in the liver is depleted in the first 24 hours, and after that the body's fat and protein are used for heat and energy. The increased use of fat without the simultaneous use of carbohydrate will lead to the formation of ketone bodies, and **ketosis** may develop [p. 69].

Tissue protein is broken down and the amino acids used to form glycogen and glucose. In addition amino acids from broken-down tissue protein are used to maintain the creatine in the muscles. About 2 per cent. of this creatine is lost each day in the urine in the form of creatinine.

Amino acids are also used for the growth of the skin, finger nails, and hair, etc. which still continues. Death

occurs in about four weeks, when the body weight has fallen to about one half.

Malnutrition

Infants and children living in the tropics on a diet abundant in starch but lacking in amino acids suffer from malnutrition even though the Calories are adequate. They have a pathetic facial expression, disfigured by skin lesions, and are bloated with oedema due to a plasma protein deficiency [p. 43]. This condition is termed **kwashiorkor.**

If the diet is also deficient in starch and Calories, the condition is termed **marasmus.** It has a very poor prognosis.

There are often signs of vitamin deficiency in these conditions.

Intravenous Feeding

A patient who is unable to take food by mouth can be fed by an intravenous infusion of glucose, amino acids and fats. Five per cent. glucose is isotonic with blood and can be used to satisfy the body's carbohydrate requirements. The protein requirements can be satisfied by giving a mixture of amino acids which are obtained by the hydrolysis of the protein casein from cow's milk. The fat requirements are satisfied by emulsions of soya bean oil or cotton seed oil. In addition mineral salts and vitamins will be required.

It is usual to give the fat emulsion separately from the carbohydrate and amino acid, and to give *heparin* for its fat clearing action (that is, to enable the fat to be removed from the blood and incorporated in the cells).

VITAMINS

In 1910 Hopkins showed that rats could not survive on a diet of pure carbohydrate, protein and fat only. But if a small amount of fresh milk were added to the diet, then the animals thrived. Hopkins postulated that an 'accessory food factor' was essential for growth, and we now know that there is a series of such substances and they are termed vitamins.

VITAMIN A

Vitamin A is a fat-soluble vitamin found in the fat of milk (butter and cream). It is absent from vegetable fats and oils (olive oil, linseed oil, ground-nut oil). Vitamin A is added to margarine during its manufacture from these oils.

Vitamin A is stored in the liver in both man and animals. Hence liver oils, particularly those from fish (cod-liver oil, halibut-liver oil), are a rich source of this vitamin.

Carotenes which are found in green vegetables and carrots, consist of two vitamin A molecules joined together. These substances are converted to one molecule of vitamin A in the body and provide an alternative source of this vitamin.

---

[5] Female readers will be able to compare this with the number of square yards of material used to make a dress. A square metre is slightly greater than a square yard.

Vitamin A is absorbed with the fat in the diet. Bile salts and pancreatic lipase are needed for this absorption.

A deficiency of vitamin A leads to changes in epithelium. The lacrimal glands, salivary glands and sweat glands degenerate. The degeneration of the lacrimal gland leads to dryness in the eyes and blindness (xerophthalmia = dry eye).

Vitamin A is also important for the formation of **visual purple** which is a pigment used by the rods of the retina for dark-adapted vision [p. 138]. A deficiency of vitamin A leads to night blindness, and the onset of this condition is often one of the earliest signs of a vitamin A deficiency.

VITAMIN B COMPLEX

The B group of vitamins are water soluble. The members of this group were originally numbered, but to prevent confusion names are frequently employed.

Many of these B vitamins are necessary for the formation of enzymes and co-enzymes which are important in the Krebs citric acid cycle and carbohydrate metabolism.

Thiamine (Aneurine — Vitamin B_1)

Thiamine is found in cereals and yeast. A deficiency of thiamine leads to **beriberi.** This is a disorder of carbohydrate metabolism which leads to polyneuritis (a disorder of both sensory and motor nerves) which is often associated with oedema and cardiac failure.

Nicotinic Acid (Niacin)

Nicotinic acid is found in liver, kidney and yeast. A deficiency of nicotinic acid gives rise to **pellagra.** Changes occur in the skin leading to dermatitis; diarrhoea is common; and the brain is affected leading to dementia. This disease is common amongst the older children and adults in the poorly nourished populations living almost entirely on maize. Infants and young children tend to develop kwashiorkor.

Riboflavin (B_2)

Riboflavin is found in whole meal flour, meat and milk. A deficiency leads to inflammation of the tongue and seborrhoeic dermatitis of the skin. Blood vessels grow into the cornea of the eye and this together with the retrobulbar neuritis leads to defective vision.

Cyanocobalamin (B_{12}) and Folic Acid

These vitamins are grouped together because they are essential for red cell formation. A deficiency of vitamin B_{12} is usually due to failure of intestinal absorption rather than a failure of intake. A deficiency leads to a macrocytic anaemia of the pernicious anaemia type. Folic acid is also needed for the formation of red blood cells, and it is possible that the vitamin B_{12} acts by making the folic acid more readily available in the bone marrow cells.

A vitamin B_{12} deficiency leads to 'vitamin B_{12} neuropathy'. This neuropathy takes the form of **subacute combined degeneration** of the spinal cord and is associated with the degeneration of both sensory and motor nerve fibres.

The macrocytic anaemia will often respond to folic acid alone but the neuropathy does not.

The adequate absorption of B_{12} by the time the end of the ileum is reached is dependent upon the secretion of the 'intrinsic factor' by the stomach. It is possible to by-pass this intrinsic factor absorption mechanism by giving large doses of B_{12} by mouth or by injecting the vitamin.

Vitamin B_{12} is stored in the liver.

Other Members of the Vitamin B Group

The other members of the vitamin B group include **pantothenic acid, pyridoxin** (vitamin B_6) and **biotin.**

Riboflavin, nicotinic acid, folic acid and biotin are synthesized by the bacteria living in the large intestine and are absorbed in significant amounts.

VITAMIN C [Ascorbic Acid]

Vitamin C is a water-soluble vitamin which is destroyed by heat in alkaline solution. It will be destroyed by cooking if soda has been added.

Vitamin C is found in citrus fruits (lemon, orange and grapefruit), blackcurrants, strawberries and leafy green vegetables. It is absent from meat, fish, fat and oils.

A deficiency of vitamin C leads to **scurvy,** which is a **haemorrhagic disease** associated with bleeding gums, haemorrhages into joints, and easy bruising of the skin.

A deficiency is most likely to occur when there is an absence of fresh fruit in the diet.

Vitamin C is probably involved in the formation of collagen and the cement substance between cells. A lack of vitamin C leads to slow wound healing.

VITAMIN D

This is a fat-soluble vitamin found in the fat of milk (butter and cream), egg yolk and fish-liver oils. Like vitamin A, it is absent from vegetable fats and oils and is added to margarine during the manufacture. Vitamin D is also made by the action of sunlight on the skin.

A deficiency of vitamin D leads to **rickets.** This is a disease of childhood associated with a decrease in the intestinal absorption of calcium and a deficiency of calcium in bones. There is excessive preparation of the cartilage for ossification, but a **deficiency of calcification.** This leads to swellings at the epiphyseal junctions. The bones are soft leading to bowing of the legs and deformation of the thorax. A similar condition which appears in adults, particularly in pregnant women, is termed **osteomalacia.**

Recent work indicates that vitamin D is converted into hydroxy-vitamin D in the liver, and then to dihydroxy-vitamin D by the kidney when the blood calcium level is low. The dihydroxy compound is more potent than vitamin D itself and greatly increases the absorption of calcium from the digestive tract in this condition.

When the plasma calcium has built up to its normal level (10 mg. per 100 ml. plasma), this conversion ceases and the calcium absorption declines.

VITAMIN E

A deficiency of this vitamin leads to sterility in rats but there is no definite evidence that it is essential for man.

VITAMIN K

This vitamin is found in green vegetables. It is needed for the formation of prothrombin [p. 39]. It is synthesized by bacteria in the intestines, and a deficiency is rare except when the intestines are sterile (at birth and after taking a broad-spectrum antibiotic).

VITAMIN DEFICIENCY

Gross vitamin deficiencies may occur in under-developed countries. These deficiencies include xerophthalmia (vitamin A deficiency), beriberi (thiamine deficiency), pellagra (nicotinic acid deficiency), scurvy (vitamin C deficiency), rickets (vitamin D deficiency). In a well-developed country a less severe form of vitamin deficiency may occur. This applies particularly to the B group where the requirements are raised by a high calorie, high carbohydrate diet, and to vitamin C in old people who are not eating fresh fruits. All the vitamins are available as chemicals and a dietary supplement may be given where necessary.

Excessive Vitamins

An excess of vitamin B and vitamin C in the diet has no apparent effect, but an excessive amount of vitamin A (hypervitaminosis A) leads to gastro-intestinal disturbances and dermatitis. The liver of some animals, such as the polar bear, contains so much vitamin A that toxic symptoms are produced if the liver is eaten as food. An excess of vitamin D is also harmful and causes a loss of weight and excessive calcification which may extend to tissues other than bone.

ESSENTIAL FATTY ACIDS

Certain unsaturated fatty acids, such as linoleic acid, are needed by the body, but cannot be synthesized by the body from other sources. They must, therefore, be present in the ingested fat. Such fatty acids are termed **essential fatty acids.** They are not classified as vitamins because they are needed in large amounts.

Prostaglandins. The unsaturated fatty acids are required for the formation of **prostaglandins.** These are very potent substances found in many tissues of the body. They have numerous actions, but appear to act mainly on smooth muscle. Prostaglandins are destroyed very rapidly in the body and therefore have a high turn-over rate.

Prostaglandins are themselves 20 carbon atom unsaturated fatty acids joined with a chemical bond between carbon atoms 8 and 12. They are subdivided into **PGA, PGB, PGC, PGD, PGE,** and **PGF** (**PG** for prostaglandin). A suffix 1, 2, 3 is then added to indicate the number of unsaturated double bonds in the molecule. Where there is more than one member the group is subdivided into α, β, γ, etc. Thus the prostaglandins which bring about uterine contraction are termed PGE_2 and $PGF_{2\alpha}$.

MINERAL SALTS AND WATER

In addition to vitamins the diet must supply the body with sufficient mineral salts.

Sodium and Potassium Salts

The dietary intake of sodium and potassium must be adequate to maintain electrolyte balance in the body. This is discussed further in the next chapter.

Calcium

Calcium is needed in the body for the ossification of bone, to maintain the correct excitability of nerve fibres, and to enable blood to clot.

Calcium is found in milk, cheese, cereals and green vegetables, but the absorption of calcium from the intestines is far from complete, and as much as three-quarters of the calcium ingested may appear in the faeces. Vitamin D is needed for adequate calcium absorption and for its utilization in the body. The blood calcium level is also regulated by the parathyroid and thyroid glands [see p. 103].

Iodine

Iodine is needed in the diet for the proper functioning of the thyroid gland. Iodine is present in sea food and in crops grown near the sea. In areas remote from the sea there may be a deficiency of iodine in the diet, and this leads to swelling of the thyroid gland (iodine deficiency goitre) and a deficiency of thyroxine in the blood [see p. 101]. Iodides are added to table salt to prevent such a deficiency.

Iron

Iron is present in most foods but is relatively deficient in milk. It is needed for the formation of haemoglobin and the iron-containing pigments and enzymes of the body.

WATER

A daily intake of water is necessary to maintain the water balance. Although a starving man can survive for several weeks without food, he can only live a few days without water.

In order to stay in water equilibrium it is necessary that the water intake each day shall be equal to the water lost. This will be discussed in the next chapter.

13. FLUID AND ELECTROLYTE BALANCE AND THE KIDNEY

The Concept of Balance

The cells of the body require a constant environment which is independent of changes in the outside world.

In order that the amount of a substance in the body may remain constant, the amount gained each day must be equal to the amount lost each day. This fact may be represented diagrammatically by a pair of scales with the intake on one side and the loss on the other side. If the intake matches the loss, the scales will be balanced, and the pointer will indicate the equilibrium position [FIG. 99]. If the intake exceeds the loss, the pointer

The kidney plays an important part in:

1. Maintenance of water balance
2. Maintenance of electrolyte balance
3. Maintenance of blood pH.

The excreted urine contains:

1. Surplus water

By excreting the water surplus to the body's requirements the kidney maintains the body in water balance [FIG. 99].

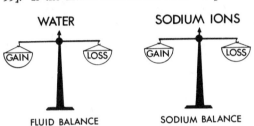
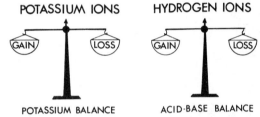

FIG. 99. Maintenance of balance by the kidney. The kidney plays an important role in the maintenance of water, sodium, potassium and hydrogen ions balance.

will indicate that the substance is accumulating in the body, leading to an excess. If the loss exceeds the intake, the pointer will indicate that the substance is being lost from the body, leading to a lack. A constant level of water and inorganic salts is required to maintain a suitable environment for the body's cells.

The amount of these substances gained each day depends on the diet, but without chemical analysis, the inorganic content of the diet is not accurately known. It is thus difficult to adjust the intake exactly to the body's requirements. Instead an excess is ingested, and balance is maintained by the regulation of the quantity excreted by the kidneys.

If, however, this excretory pathway fails, as for example, in renal failure [p. 86], the intake must be carefully adjusted to be exactly equal to the losses by pathways other than the kidneys (sweat, faeces, etc.). An unlimited intake such as is normally taken would lead to an accumulation of the substance in the body.

THE KIDNEY

The principal function of the kidney is to produce urine. It also produces renin [pp. 22, 44], erythropoietin [p. 35] and converts vitamin D to an active form which promotes the absorption of calcium from the digestive tract [p. 74].

By producing urine which is excreted from the body, the kidney maintains the constancy of the *internal environment*, that is, it **maintains the body fluids at a constant composition**.

2. Surplus electrolytes

By excreting the ingested inorganic ions which are surplus to the body's requirements (sodium, potassium, magnesium, calcium, chloride, etc.) the kidney maintains the body in electrolyte balance. Clinically the most important are sodium ions, potassium ions and hydrogen ions [FIG. 99].

3. Metabolic waste products

(a) **Urea**

The urea is formed from the amino acids which are used for heat and energy. Before this can occur the nitrogen part of the molecule is converted to ammonia and then to urea [p. 66].

(b) **Uric acid**

The nitrogen from nucleic acids[6] and purines is excreted in the form of uric acid. Some uric acid is synthesized in the body. An excessive production of this substance and its deposition in joints leads to the painful condition of gout.

(c) **Creatinine**

The creatinine in the urine comes from the creatine in muscle, and its presence in urine represents a loss of nitrogen from the body. Creatine phosphate is a source

---

6 Nucleic acids are found in the nuclei of cells. They are therefore found in meat and fish protein (mainly muscle cells) but will be absent from milk protein (no cells in milk).

of heat and energy in muscle. The creatine is continually breaking down into creatinine which leaks into the blood and is lost in the urine. This loss, which is related to the amount of muscle tissue, places a great strain on the protein resources of the body in starvation, since this loss will continue even though there is no food protein intake.

4. Surplus acids and surplus alkalis

The urine is the route of excretion of surplus acids (or alkalis) that have been ingested or made by metabolism (metabolic acidosis). The pH of the urine indicates whether the surplus is one of acids or one of alkalis [see also p. 84].

5. Abnormal constituents

The presence of abnormal constituents in the urine often gives a clue to the underlying disorder. Some of the abnormal constituents for which tests are made are (with the disorder in brackets):

1. Glucose (diabetes mellitus) [p. 70].
2. Ketone bodies (ketosis) [p. 69].
3. Plasma albumin (kidney disease).
4. Red cells (kidney disease).
5. Galactose (galactosaemia) [p. 67].
6. Phenylketones (phenylketonuria) [p. 67].

The urine will also contain the metabolic products of all drugs that have been taken, and will probably contain traces of all the constituents of the blood. The presence, for example, of the hormone **chorionic gonadotrophin** in the urine of a pregnant woman forms the basis of pregnancy tests [p. 110].

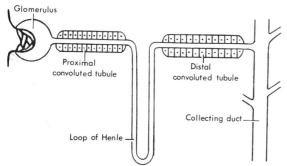

FIG. 100. Diagram showing the constituent parts of the nephron (not to scale).

THE FORMATION OF URINE

The fundamental unit for the formation of urine is the **nephron,** and there are one million nephrons in each kidney. Each nephron starts with a tuft of six to eight blood capillaries invaginated into the end of a tubule. This structure is called a **glomerulus.** The tubule is divided into the **proximal convoluted tubule,** the **loop of Henle** and the **distal convoluted tubule** [FIG. 100]. Several tubules lead to each **collecting duct** which leads to the bladder via the **ureter.** The urine is voided from the **bladder** via the **urethra.**

Urine is formed by the process of:
1. **Glomerular filtration**
2. **Tubular reabsorption**
3. **Tubular secretion**

Glomerular Filtration

The glomerulus acts as a filter between the blood and the tubule. About one tenth of the water in the blood flowing through the glomerulus, and its associated small constituents (molecular weight less than 67,000), are filtered and pass into the tubule. The large constituents, including the blood cells and plasma proteins, are not filtered and remain in the glomerular capillaries [TABLE 3].

TABLE 3
Glomerular filtration

Large constituents and molecules Not filtered	Small molecules Filtered and enter tubule
Blood cells Plasma proteins	Water Inorganic salts Glucose Amino acids Urea Uric acid Creatinine } waste products

The filtrate passing into the tubule will contain the inorganic salts, food substances such as glucose and amino acids, and the waste products, urea, uric acid and creatinine.

This process of filtration is entirely non-selective. The substances filtered are those of small size and there is no modification at this stage whatever the body's requirements.

Tubular Reabsorption

The blood capillaries of the glomerulus form into a second arteriole which runs to the tubule. Here it forms a second capillary network surrounding the tubule. This second capillary network allows **reabsorption** to take place [FIG. 101]. This reabsorption is **selective** and varies according to the body's needs for each substance. If there is a shortage in the body then complete reabsorption occurs. If there is a surplus of a substance, the excess is allowed to continue down the tubule without reabsorption and to pass to the collecting ducts, ureter and the bladder.

Glucose is an example of a substance that is reabsorbed completely under normal conditions.

Tubular Secretion

The third process involved in the production of urine is tubular **secretion.** Substances are added to the glomerular filtrate by active transport in the reverse direction to reabsorption. Surplus acids and alkalis are removed from the body by this secretory mechanism.

Penicillin is an example of a substance which is not only filtered by the glomerulus, but is also lost by tubular secretion. The drug *probenecid*, which inhibits the tubular secretion of penicillin, is given to raise the mean blood level when patients are receiving penicillin.

A study of kidney function is thus principally a study of the modification to the glomerular filtrate which occurs in the tubule. Since each substance is treated differently, the principal constituents of the glomerular filtrate will be considered in turn.

WATER

All the nephrons working together filter 120 ml. water per minute. This is known as the **glomerular**

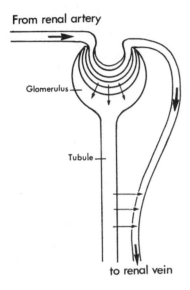

From renal artery

Glomerulus

Tubule

to renal vein

FIG. 101. The capillaries of the glomerulus reform into an arteriole which leads to a second capillary network around the tubule. This allows glomerular filtration to be followed by tubular reabsorption.

filtration rate (GFR). It is equivalent to the filtration of 170 litres water per day.

This is represented diagrammatically in FIGURE 102 where one large theoretical nephron represents all the nephrons in the kidneys.

Of the 120 ml. water filtered each minute about 119 ml. are reabsorbed as the filtrate passes along the tubule, and only 1 ml. passes to the collecting duct, ureter, bladder and becomes urine. 1 ml. per minute will amount to 1,500 ml. urine per day.

It will be noted that the blood only contains 3 litres of water in the plasma, and a glomerular filtration of 170 litres of water per day means that the same water will be filtered and reabsorbed 50 to 60 times a day (2–3 times each hour). There will also be an interchange between the water in the plasma and the water in the rest of the body (cells and tissue spaces). But the total body water is only 45 litres, and the 170 litres per day

glomerular filtration would mean that all the water in the body will be filtered on the average four times every day.

Tubular Reabsorption of Water

The tubules are extremely long tubes (many centimetres in length) when compared with a red blood cell (7 microns diameter) and the size of the glomerulus (100 microns). To give some idea of this length, if the glomerulus were increased in size to that of a golf ball then a typical tubule would be 30 feet long (9 metres).

During the passage along the extremely long tubule, the constituents to be reabsorbed enter on one side of the cells lining the tubule, pass through these cells, and enter the blood capillaries on the other side. In the case of most substances this is an active process on the part of the tubular cells and requires the expenditure of energy.

As the filtrate passes along the proximal convoluted tubule, seven-eighths of the sodium chloride is absorbed by this active process and seven-eighths of the water is reabsorbed passively along with this sodium.

The reabsorption of the remaining water occurs in the distal convoluted tubules and collecting ducts and is controlled by the antidiuretic hormone (ADH) from the posterior pituitary gland [p. 99]. ADH is also known as vasopressin. ADH release is itself controlled by **osmoreceptors** in the region of the hypothalamus.

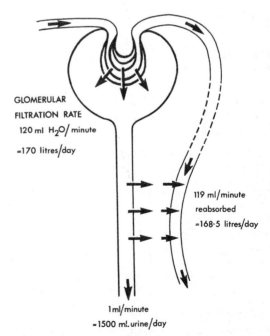

TOTAL BODY WATER = 45 LITRES

PLASMA VOLUME = 3 LITRES

GLOMERULAR FILTRATION RATE

120 ml H_2O/ minute

=170 litres/day

119 ml/minute reabsorbed =168·5 litres/day

1 ml/minute =1500 ml. urine/day

FIG. 102. Diagram to show water turn-over in the nephron. The 'giant' nephron represents all the 2,000,000 nephrons in the two kidneys.

MAINTENANCE OF WATER BALANCE BY THE KIDNEYS

Fluid Compartments

The total water in the body amounts to about 45 litres (**total body water**). About 30 litres of this is found inside the cells (**intracellular fluid**) whilst the remaining 15 litres is outside the cells (**extracellular fluid**). The 15 litres extracellular fluid is subdivided into 12 litres **tissue fluid** [p. 41] and 3 litres **plasma** [FIG. 103].

Water makes up approximately 70 per cent. of the body weight. Since the body weight remains constant, it implies that the total water in the body is constant.

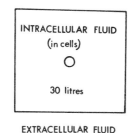

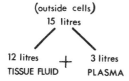

FIG. 103. Body water compartments.

For the body to remain in water balance, the water gained each day must be equal to the water lost. If the water gained exceeds the water lost, then the subject will become over-hydrated and oedematous. On the other hand, if the water lost exceeds the water gained, then dehydration will ensue.

Water Gained

Water is gained each day in several ways [FIG. 104]. Firstly, fluids are taken in by mouth. In this connexion it should be remembered that a cup of tea is equal to a cup of water. A cup of coffee is equal to a cup of water, a glass of milk is equal to a glass of water, etc.

Secondly, water is taken in as moisture in the food. Even the driest biscuit contains quite a high percentage of water, and some fruits such as the melon contain well over 90 per cent. of water.

Thirdly, water is made in the cells of the body by the oxidation of hydrogen in the food during the metabolic processes which produce heat and energy [p. 45]. Each day about half a litre of water will be made in this way.

The intake of water is safeguarded by the sensation of **thirst,** but the intake of fluids may be determined by other factors, such as social habits, rather than the body's need for water. Swilling out the mouth with

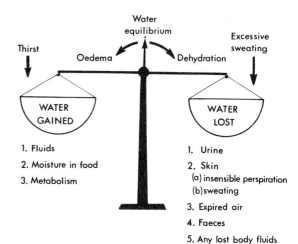

FIG. 104. Factors maintaining water balance. If the water gained equals the water lost each day the body will be in water equilibrium. If the water gained is less than the water lost, dehydration ensues. If the water gained exceeds the water lost, the body becomes over-hydrated leading to oedema. The sensation of thirst ensures that the water intake is adequate.

water will temporarily quench the thirst although no water is swallowed.

Water Lost

Water is lost in the urine, through the skin, in the expired air, in the faeces, and in any lost body fluids.

Skin. Water is lost through the skin in two ways. The skin is not a completely water-tight covering to the body and some water is continually being lost from the tissue spaces by the process of **insensible perspiration.**

This loss of water from the surface of the body is normally very small but it may be very large following burns in which vast areas of skin have been lost.

When the body is hot, **sweating** occurs. The sweat, which is a dilute solution of sodium chloride in water, reaches the skin via the sweat-gland ducts. Sweating is brought about by the activity of the sympathetic nervous system, but the sympathetic nerve fibres to the sweat glands release acetylcholine (instead of the usual sympathetic transmitter noradrenaline). The evaporation of this sweat cools the skin [p. 88].

Expired Air. Expired air is fully saturated with water vapour. This fact is used as the basis of the 'mirror test' to see if a patient is breathing, or if leaks are occurring around a respiratory spirometer mouth-piece during a lung function test. The water vapour makes a shiny mirror cloudy.

Since the inspired air contains less water vapour than the expired air, the process of breathing causes a loss of water from the body.

The loss of water in the expired air can be high in cases of increased respiration, particularly if the inspired air has a low water content. Thus a patient who is hyperventilating will be losing more water in this way than a normal person.

The loss of water in expired air can be very high in mountaineering, since the cold environmental temperature makes the inspired air very dry, whilst the high altitudes will increase breathing. At very high altitudes with the intense cold, the provision of fuel to melt the ice in order to provide drinking water presents a serious problem.

It should be remembered that the introduction of central heating into a building may increase the water loss through the skin and in the expired air of the occupants and unless the fluid intake is increased, this could lead to dehydration particularly in elderly people.

Faeces. The amount of water lost in the faeces is, under normal conditions, quite small (150 ml.). But in diarrhoea the water lost this way may be very high and this loss may lead to severe dehydration.

Water Lost in other Body Fluids. Water will be lost from the body by vomiting and any haemorrhage represents a loss of that volume of water. Even the common cold may cause a substantial loss of water from the nose. Crying also represents a loss of water.

Urine. Very few of the above factors affecting water intake and water loss can be altered in order to maintain the water balance. Under normal conditions the only variable which enables water equilibrium to be maintained is the volume of urine. Furthermore the fluid intake is frequently determined by social considerations rather than the physiological needs. Thus if one accepts an additional cup of coffee or tea, the body will have an increased intake of 100 ml. of water, which must then be excreted in the urine to maintain the water balance.

Water Excess

The mechanism whereby water equilibrium is maintained may be seen by considering the effects of drinking a litre of water surplus to the body's needs. The water is taken by mouth, it passes through the stomach, and

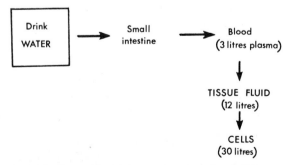

FIG. 105. Water taken in by mouth is absorbed into the blood in the small intestine. It is then shared between the water compartments of the body (blood, tissue fluid, cells) in proportion to their total volume.

is absorbed into the blood in the small intestine. Here it dilutes the plasma, but the majority of water leaves the plasma and passes into tissue fluid and then into the cells of the body [FIG. 105].

The dilution of the blood affects the osmoreceptors in the region of the hypothalamus and the amount of posterior pituitary hormone ADH (antidiuretic hormone) is reduced. This hormone normally facilitates the reabsorption of water in the distal tubules. As a result of the reduction in the ADH level, less water will be reabsorbed and there will be a diuresis (increase in urine volume). This increase in the volume of urine will continue until the water balance has been restored to normal.

Rate of Tubular Reabsorption of Water

The glomerular filtration rate is a constant 120 ml./minute. The tubular reabsorption under the influence of a normal level of ADH is initially 119 ml./minute. Thus 1 ml. of urine/minute passes to the bladder. At the height of the diuresis the ADH level will fall almost to zero and the tubular reabsorption will fall to 105 ml./minute [TABLE 4].

It does not fall below 105 ml./minute because seven-eighths of the water is reabsorbed in the proximal tubule and this reabsorption is not under the control of ADH. ($\frac{7}{8}$ × 120 ml./minute = 105 ml./minute).

TABLE 4

	GFR	Tubular reabsorption	Urine per minute	Urine per day
Hot day, sweating, ADH level high	120 ml./min.	$119\frac{3}{4}$ ml./min.	$\frac{1}{4}$ ml.	375 ml.
Normal	120 ml./min.	119 ml./min.	1 ml.	1,500 ml.
No ADH	120 ml./min.	105 ml./min.	15 ml.	22,500 ml.

Diabetes Insipidus

The condition of **diabetes insipidus** is a condition in which the posterior pituitary gland fails to produce ADH. As a result a very large volume of urine is produced each day (it may be in excess of 20 litres), and the subject becomes extremely thirsty.

The term *insipidus* dates from the time when the only method of testing urine was to taste it, and in this condition the urine was so dilute, that it was tasteless or insipid [c.f. diabetes mellitus, p. 70].

Water Lack

On a hot summer's day when a large amount of water is being lost by sweating, the ADH level is increased, the tubular reabsorption increases to $119\frac{3}{4}$ ml./minute, and the volume of urine produced drops to $\frac{1}{4}$ ml./minute and the urine becomes very concentrated. This volume is equivalent to 375 ml./day, which is the minimum obligatory volume for the excretion of the urea and other waste products.

Treatment of Dehydration

If a water deficit has occurred, leading to dehydration, it may be treated by giving water by mouth. If this is not possible, water containing 5 per cent. glucose, to give it the same osmotic pressure as blood, may be given by intravenous injection, or by a tube inserted into the stomach or rectum.

If there is no sodium depletion, isotonic saline should not be used because this would overload the body with sodium.

Osmotic Diuresis

If the urine contains a large quantity of excretory products, such as glucose, or an excessive amount of urea, then the osmotic pressure exerted by these substances will prevent the normal reabsorption of water, and the volume of urine will be increased. This is termed an **osmotic diuresis**. Potassium citrate is an example of a substance which acts as a diuretic in this manner. Another is the sugar **mannitol**.

Other **diuretics** such as *mersalyl* and *chlorothiazide* act by inhibiting the reabsorption of sodium. The water that could have been passively reabsorbed with this sodium is now excreted [p. 82].

Weight Reduction by Dieting and Water Loss

Any sudden change in body weight can only be brought about by a change in the amount of body water.

It is interesting to note that although 1 stone (14 lbs.) in weight can be lost in a matter of a few hours by profuse sweating in a Turkish bath, it is difficult to lose anything like this amount of weight in a short time by dieting.

Consider the theoretical case of a person who is eating no food at all and is living entirely on body fat. Each gram of fat gives 9 Calories. If the daily requirements are 2,700 Calories per day, this could be provided by 300 grams (= 10 oz.) of fat. The loss of body fat will thus be 10 oz. per day or 70 oz. (2·1 kg.) per week. This weight loss of 2·1 kg. is equivalent to only 4½ lb. per week, and is assuming complete starvation with no food intake whatsoever. However, a weight loss by dieting of only a few ounces per day, will amount to a significant loss of weight over a period of a year.

Over-hydration. Too much water in the body leads to mental changes which include nausea, anorexia, vomiting and finally confusion, stupor and epileptiform attacks.

Under-hydration. Under-hydration (dehydration) leads to thirst and finally to confusion and coma particularly in the elderly.

MAINTENANCE OF ELECTROLYTE BALANCE

By excreting the surplus of inorganic substances, the kidney maintains the electrolyte balance of the body. Most of the sodium, potassium and magnesium salts in the body fluids are completely dissociated and exist in solution as electrically charged ions. Thus sodium chloride in solution exists as sodium ions Na^+ (with one electron less than a sodium atom) and chloride ions Cl^- (with one electron more than a chloride atom). 'Sodium' in body fluids is short for 'sodium ions' and not metallic sodium.

Glucose and urea, which are not electrolytes, are not ionized in solution.

Millimoles per Litre

The body needs a constant internal environment. As far as the cells are concerned, it appears that it is the **concentration** of each electrolyte in the body fluids that is important rather than the total amount of the electrolyte in the body. In other words, the cells need the correct number of molecules per litre of fluid. Expressing a concentration in **milligrams per litre** does not give any indication of the number of molecules present since some molecules are heavier than others. To allow for this difference, the concentrations are expressed in *milligrams divided by the molecular weight* per litre, and these units are called **millimoles per litre** (mmol./1.).

Molecular weights and ionic weights are the sum of the atomic weights of the constituent atoms. The electron is so light that no allowance need be made for the addition of, or lack of, electrons in the ions.

ATOMIC WEIGHTS

H	1	Mg	24
C	12	Cl	35·5
N	14	S	32
O	16	K	39
Na	23	Ca	40

Isotonic (Physiological) Saline

This is made by dissolving 0·9 grams of sodium chloride in every 100 ml. water (0·9 per cent w/v). This is equal to 9000 milligrams of NaCl per litre.

The atomic weight of sodium is 23 and that of chlorine 35·5, so that the molecular weight of sodium chloride is $23 + 35·5 = 58·5$.

The concentration of sodium chloride in isotonic saline is thus 9000/58·5 millimoles per litre = **154 mmol./l.**

It contains: 154 mmol. per litre of sodium ions and 154 mmol. per litre of chloride ions.

Milliequivalents per Litre

If the concentration of an ion in millimoles per litre is multipled by the valency of the ion, the resultant concentration is expressed in **milliequivalents per litre.** Fortunately, the majority of the ions encountered in body fluids, hydrogen ions, sodium ions, potassium ions, chloride ions and bicarbonate ions all have a valency of 1. As a result the numerical value of their concentration in *millimoles per litre* is the same as their concentrations in *milliequivalents per litre*. (Valency is the number of + or − signs shown against the ion.)

Thus isotonic saline contains:

154 milliequivalents per litre of sodium ions (Na^+).

154 milliequivalents per litre of chloride ions (Cl^-).

However, Mg^{2+} and Ca^{2+} have a valency of 2. Thus 24 mg. per litre of Mg^{2+} equals 1 millimole per litre of Mg^{2+} but 2 milliequivalents per litre of Mg^{2+}.

Milliequivalents per litre are obsolescent units since it has been proposed that only *millimoles per litre* be used when the S.I. units are universally adopted in the not too distant future.

Extracellular Fluid

The principal ions found in the extracellular fluid are **sodium** ions (Na^+) which are positively charged, and chloride ions (Cl^-) and bicarbonate ions (HCO_3^-) which are negatively charged [FIG. 106].

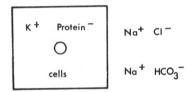

Extracellular fluid

Sodium salts in extracellular fluid

Potassium salts in intracellular fluid

FIG. 106. The principal ions found in the intracellular and extracellular fluids.

The plasma differs from the other extracellular fluid, tissue fluid, in that it contains protein (plasma proteins) which at a pH of 7·4 will be negatively charged ions. Plasma thus has sodium as its main positively charged ion; with chloride, bicarbonate and protein as its negatively charged ions.

If the quantities of these ions are expressed in milliequivalents per litre then the equality between the numbers of positively charged ions and the negatively charged ions in plasma may be represented graphically [FIG. 107].

The concentration of sodium (Na^+) ions is 145 milliequivalents per litre whilst that of the chloride (Cl^-) ions is 105 milliequivalents per litre, that of bicarbonate ions (HCO_3^-) is 26 milliequivalents per litre, that of protein ions (Pr^-) is 14 milliequivalents per litre.

$$145 = 105 + 26 + 14$$

Intracellular Fluid

The main positively charged ions found in cells are **potassium** and magnesium, whilst the main negatively

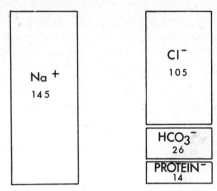

MILLIEQUIVALENTS PER LITRE

FIG. 107. Concentration of ions in plasma. If these concentrations are expressed in milliequivalents per litre, the total of the positive ion concentration will equal the total of the negative ion concentration (145 = 105 + 26 + 14).

charged ions are protein, phosphate and sulphate. The concentration of potassium in cells is similar to the concentration of sodium in the extracellular fluid, namely 150 milliequivalents per litre.

It will be noted, however, that although sodium and potassium have very similar chemical properties, in the body they are entirely different substances and are in no way interchangeable.

SODIUM

The principal sodium salt in the body is **sodium chloride** and this is the only inorganic chemical that is eaten as such in the diet (common salt). All the other inorganic requirements of the body are satisfied by the presence of these substances in the ingested food.

Sodium Balance

A balance is maintained between the sodium chloride gained each day and the sodium chloride lost [FIG. 108].

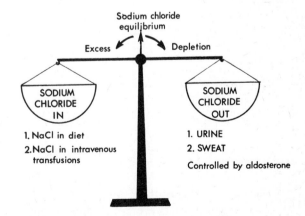

FIG. 108. The factors regulating the salt balance of the body.

Sodium chloride is gained by the body by the ingestion of common salt. Sodium chloride will also be gained by the body when an intravenous transfusion of isotonic saline is given.

Sodium chloride is lost by the body in the urine and in sweat.

Under normal conditions, the sodium chloride intake exceeds that lost by sweating and the surplus is excreted in the urine. There is an enormous turnover of sodium chloride in the kidney. A daily glomerular filtration of 170 litres of water containing 0·9 per cent. NaCl means that 1,500 g. of sodium chloride are filtered. If 15 g. appear in the urine, 1,485 g. must be reabsorbed each day!

Regulation of Sodium Reabsorption

The maintenance of sodium balance is under the control of the hormone **aldosterone** from the adrenal cortex. Aldosterone increases the reabsorption of sodium by the kidney tubules and reduces the sodium content of sweat.

Any sodium depletion will lead to an increase in the circulating aldosterone, which will, in turn, lead to the retention of sodium by the kidneys. The controlling mechanism for the increased release of aldosterone is not fully understood but it is probably a result of an increased production of **renin** by the kidney, which is followed by the formation of **angiotensin** [p. 22]. Angiotensin stimulates the release of **aldosterone** from the adrenal cortex.

In severe sodium deprivation, sodium chloride may disappear from the urine completely since it is being totally reabsorbed.

Sodium Lack (Hyponatraemia)

If an excessive amount of water is ingested, the sodium in the body will be diluted, and this leads to sodium lack, or hyponatraemia (*natrium* (L) = sodium).

It is the concentration of the electrolytes in the water that is important and not the total quantity of electrolytes in the body. Any increase in body water will reduce the concentration of the sodium and potassium.

The effects of ingestion of an excessive amount of water are termed **water intoxication.** It leads to cell oedema, headaches, vomiting and painful cramps.

The fact that a sodium deficiency would lead to painful muscle cramps was first shown in stokers by Haldane, and is referred to as stoker's cramp. Stokers work in a very hot environment and sweat profusely. Sodium lack is the result of the excessive sweating. Sodium chloride is lost in the sweat and only the water is replaced by drinking.

The treatment of the painful muscle cramp is to take salt tablets as well as the water.

Sodium Excess (Hypernatraemia)

The excess of sodium is called hypernatraemia, and is associated with insufficient water to match the sodium in the body.

The effects are thirst, a dryness of the mucous membranes, fever and nervous irritability.

The skin becomes withered, and the dehydration leads to circulatory shock and a decrease in the volume of urine.

POTASSIUM

Potassium is the main positively charged ion in cells (150 milliequivalents/litre).

The amount of **potassium** in the **plasma** is very small. Its level is 5 milliequivalents per litre. This represents the potassium in transit from the digestive tract through to the cells where it is stored.

Although this level of 5 milliequivalents/litre of potassium is very low, its maintenance is of great importance, and life is only possible if the level is maintained between 3 and 8 milliequivalents/litre.

Reabsorption of Potassium

The excess potassium is excreted in the urine. The mechanism is complex. It is probable that all the potassium in the glomerular filtrate is reabsorbed in the proximal tubules and that some potassium is secreted into the urine in the distal part of the tubules.

The secreting mechanism for potassium appears to be shared with that for the secretion of hydrogen ions from surplus acids. If the potassium to be secreted is high, it will prevent the removal of surplus acids from the body. These acids will be retained, and this combination of a high blood potassium, with a high blood acid content, is termed **hyperkalaemic metabolic acidosis** (*kalium* (L) = potassium).

If the blood potassium is low, then very little potassium will be secreted into the filtrate. Instead there will be an excessive secretion of acids by the kidney tubules into the urine. The depletion of acid in the body associated with a low potassium is termed **hypokalaemic metabolic alkalosis.**

Potassium Lack. Potassium lack may result from the excessive use of diuretics, from vomiting and from laxatives. The effects are complex and include muscle weakness, fatigue, gastric distension, polyuria and paralysis.

Unlike sodium ions which are completely reabsorbed when the plasma sodium level is low, potassium ions do not disappear from the urine when the plasma potassium level is low. As a result the potassium loss continues, and the condition of the patient deteriorates until the potassium intake is raised.

A low plasma potassium makes a patient unduly sensitive to digitalis.

DIURETICS

Oedema fluid is essentially an isotonic sodium chloride solution. One method of combating oedema is to give the patient a *diuretic*. These substances reduce the reabsorption of sodium (and hence water) by the kidney and by so doing increase the sodium and water lost in the urine.

Mersalyl, which is given by intramuscular injection, is an organic mercury compound. It acts on the proximal convoluted tubule and reduces its ability to reabsorb sodium, chloride and water.

Benzothiadiazine diuretics (which include *chlorothiazide* and *hydrochlorothiazide*) act in a similar manner, but have the advantage that they can be taken by mouth. Since less sodium is reabsorbed in the proximal tubule, the concentration of sodium in the distal tubule will be higher. This will favour the excretion of potassium for some of this sodium in a distal tubule, and lead to potassium depletion unless an **oral potassium supplement** is given.

Spironolactone is a diuretic which acts by antagonizing aldosterone. It will therefore decrease the sodium reabsorption in the distal tubule, increase the sodium lost and decrease the potassium lost.

Other diuretics such as **triamterine** act by preventing the exchange of potassium for sodium in the distal tubule.

MAINTENANCE OF BLOOD pH

Although urine is made by the glomerular filtration of **alkaline blood** (pH = 7·4), the urine reaching the collecting tubules is usually **acid**. This is because hydrogen ions have been added to the filtrate as it passes along the tubule by the process of **tubular secretion**.

On a diet containing adequate protein there is a continual production of sulphuric acid. This is because the amino acids cysteine and methionine contain sulphur in addition to carbon, hydrogen, nitrogen and oxygen. This sulphur is oxidized in the body to sulphuric acid.

The sulphuric acid is temporarily buffered by the sodium bicarbonate in the plasma, and respiration is stimulated in order to maintain the blood pH at its correct level of 7·4 [see p. 53]. But the ultimate correction takes place in the kidney where the sulphuric acid is excreted.

The sulphuric acid is excreted in buffered form. The hydrogen ions and the sulphate ions associated with the sulphuric acid are secreted by the kidney tubular cells into the filtrate. The hydrogen ions exchange for sodium ions and convert the disodium hydrogen phosphate to sodium dihydrogen phosphate. The sodium ions and the sulphate ions form sodium sulphate.

$$\text{DISODIUM HYDROGEN PHOSPHATE} + \text{SULPHURIC ACID} \rightarrow \text{SODIUM DIHYDROGEN PHOSPHATE} + \text{SODIUM SULPHATE}$$

The sulphuric acid is thus excreted as sodium sulphate and sodium dihydrogen phosphate.

A sodium dihydrogen phosphate solution has a pH of 4·5 whilst a disodium hydrogen phosphate solution has a pH of 8·5. When both are present the pH lies between these two limits [FIG. 109]. They form a buffer system which allows the acids to be excreted without the urine pH falling below pH 4·5 [FIG. 109].

The sulphuric acid is also excreted as ammonium sulphate. The ammonia is made from the amino acid glutamine by the kidney tubule cells.

Although the urine may have any pH between 4·5 and 8·5, on a normal mixed diet the pH of urine is usually in the range of 5·3 − 7.

Alkaline Urine

On a vegetarian diet, metabolic processes may lead to a surplus of alkali in the body instead of a surplus of acid (see citrate below). In such a case, hydroxyl ions, not hydrogen ions, will be secreted by the kidney tubules into the filtrate making the urine alkaline. Sodium dihydrogen phosphate will be converted into disodium hydrogen phosphate. The surplus alkali will be excreted in this form and as sodium bicarbonate.

Ions that Vanish

Although the kidneys are the route of excretion for the surplus inorganic salts taken in by mouth, some ingested ions appear to vanish and do not appear in the urine.

When sodium citrate, or sodium tartrate, is ingested (present in citrus fruits), the **citrate** and **tartrate** ions are converted by metabolism to bicarbonate, carbon dioxide and water. Sodium bicarbonate is formed and the carbon dioxide is given off in the expired air.

The ingestion of sodium citrate and sodium tartrate is equivalent to taking sodium bicarbonate, and will lead to **alkalosis** (= a tendency towards alkalaemia).

Lactate ions are also converted to bicarbonate by metabolism. Sodium lactate is more heat stable and therefore easier to sterilize than sodium bicarbonate (which tends to give off carbon dioxide and change to highly alkaline sodium carbonate—'washing soda'). For this reason sodium lactate is frequently employed in place of sodium bicarbonate in infusion fluids.

The **ammonium** ion is an example of a positively charged ion that can vanish. It is converted to urea in the body. This neutral unionized compound urea is

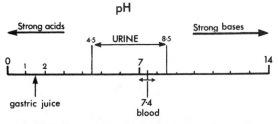

FIG. 109. By excreting the surplus acids or the surplus alkalis, the kidney maintains the blood pH at 7·4. The urine is buffered, and this enables acids and alkalis to be excreted with a urinary pH range restricted to 4·5 to 8·5.

excreted in the urine. Thus ingestion of ammonium chloride leads to **acidosis** as a result of its conversion to urea and hydrochloric acid.

Saline Infusion

It should be noted that isotonic saline has a pH of 7 and is relatively acid compared with blood pH 7·4. If large quantities of intravenous fluid are being given some of the sodium chloride should be replaced by alkaline sodium bicarbonate (or sodium lactate which will be converted to sodium bicarbonate in the body).

DERANGEMENTS OF ACID-BASE BALANCE

Respiration and Kidney in Acidaemia and Alkalaemia

A detailed investigation of a case of acidaemia or alkalaemia involves the determination of the amount of carbon dioxide in solution in the plasma and a determination of the plasma bicarbonate level. The normal values are: CO_2 in solution 3 ml. CO_2/100 ml. blood (1·3 mEq. per litre), plasma bicarbonate 26 mEq. per litre (which will be carrying 60 ml. CO_2/100 ml. blood).

If the quantity of carbon dioxide carried in the plasma as sodium bicarbonate is 20 times the carbon dioxide in solution, then the plasma pH will be at the correct level of 7·4 [p. 55].

FIGURE 110 represents this diagrammatically. A balance has one arm twenty times longer than the other. If we consider the carbon dioxide as bicarbonate attached to the shorter arm, and the carbon dioxide in solution attached to the longer arm, then the quantity of the carbon dioxide in solution will be magnified 20 times and the two sides will balance. A pointer attached at the fulcrum will then point to 7·4.

Such a balance can be upset in two ways. One side can become lighter or the other side can become heavier.

Changes which occur due primarily to changes in the bicarbonate concentration are termed **metabolic.**

Changes which occur due primarily to changes in the CO_2 in solution are termed **respiratory.**

The sodium bicarbonate in the plasma is determined by the diet and by the formation or excretion of sodium bicarbonate by the kidneys. **Standard bicarbonate** is the amount measured under standardized conditions of gas tensions (P_{CO_2} = 40 mm. Hg, P_{O_2} = 100 mm. Hg).

The CO_2 in solution is determined by the percentage of carbon dioxide in the alveolar air and this is determined by the respiration.

pH Fall

Should the pointer [FIG. 110] be deflected to the right, this will represent a fall in pH, that is **acidaemia** (or acidosis). Such a change can be brought about either by a reduction in the sodium bicarbonate level (**metabolic acidaemia**) or by an increase in the CO_2 in solution (**respiratory acidaemia**).

Metabolic acidaemia occurs when acids enter the blood. Such acids can be sulphuric acid and phosphoric

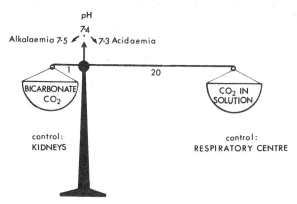

FIG. 110. If the carbon dioxide which is carried in the blood as bicarbonate is 20 times the carbon dioxide in solution the blood pH will be 7·4. This may be represented diagrammatically by a balance which has one arm 20 times the length of the other. If the balance is upset by an alteration in the bicarbonate level the resultant acidaemia or alkalaemia is termed metabolic. If the balance is upset by an alteration of the CO_2 in solution level the resultant acidaemia or alkalaemia is termed respiratory.

An increase in blood pH is termed alkalaemia. A decrease in blood pH is termed acidaemia.

acid from soft drinks, or metabolic acid such as lactic acid in exercise or acetoacetic acid in untreated diabetes mellitus.

Acidaemia, that is a fall of pH in blood, stimulates respiration via the respiratory centre, and this respiratory stimulation will lead to a reduction in the CO_2 in solution. In metabolic acidaemia the initial stages will be the reduction in bicarbonate, but the result of acidaemia will stimulate the respiratory centre and bring about a reduction in the CO_2 in solution. A new equilibrium is established with the pH only slightly reduced from the normal value of 7·4, but with both the bicarbonate and the CO_2 in solution reduced. This is termed a **compensated acidaemia.** It is detected by estimating the bicarbonate level in the plasma. This level will be low. The surplus acids are ultimately excreted in the urine.

An increase of CO_2 in solution occurs in underventilation when CO_2 is retained in the body. It can be brought about experimentally by breathing a gas mixture containing a high percentage of carbon dioxide. Normally room air contains no carbon dioxide. Breathing 5 per cent. CO_2 or 7 per cent. CO_2 in air mixture will lead to **respiratory acidaemia.**

pH Rise

Should the pointer be deflected to the left, and the pH rise, this condition is termed **alkalaemia.** It may be due to an increase in the bicarbonate in the plasma or to a reduction in the CO_2-in-solution level in the plasma.

An increase in sodium bicarbonate in the plasma follows the ingestion of sodium bicarbonate or sodium citrate and tartrate which are converted to sodium bicarbonate by metabolism [p. 83]. It also occurs in vomiting when gastric juice is lost. The hydrochloric acid in the juice is made in the stomach by the conversion of sodium chloride into hydrochloric acid and sodium bicarbonate. Normally the hydrochloric acid secreted into the stomach is neutralized by the sodium bicarbonate secreted in the pancreatic juice, thus the fluids entering the jejunum are neutral. Should the gastric juice be lost by vomiting, this will represent a loss of hydrochloric acid and the body will then have a surplus of sodium bicarbonate.

Carbon dioxide in solution is reduced when the percentage of carbon dioxide in the lungs falls. No gas mixture will bring this about since room air already contains no carbon dioxide, and no gas mixture can contain any less. The carbon dioxide in the lungs is reduced in over-breathing, and this leads to alkalaemia.

In alkalaemia the respiratory centre is depressed and as a result carbon dioxide is retained in the blood. The CO_2 in solution rises, and the new equilibrium is established with an increase in both the bicarbonate and the CO_2 in solution. This will be associated with an increased bicarbonate level.

Alkalaemia due to an excess of bicarbonate is termed **metabolic alkalaemia** whereas that due to a reduction in the CO_2 in solution is termed a **respiratory alkalaemia.**

In both acidaemia and alkalaemia the immediate correction is respiratory; the long-term correction is renal.

TUBULAR REABSORPTION OF GLUCOSE

The normal blood glucose level is 60–100 mg. glucose per 100 ml. blood. The glomerular filtrate will have the same composition. At this glucose concentration, the tubule cells are able to transfer all the glucose back into the blood by active transport, and no glucose passes to the bladder.

But the tubule cells are limited in the quantity of glucose that they can transport back to the blood in a given time (tubular maximum for glucose). When the blood glucose level exceeds 180 mg. glucose/100 ml. blood, the tubule cells are not able to reabsorb all the glucose. Some glucose reaches the bladder and **glycosuria** results.

Urine Tests for Glucose

Glucose is a reducing agent which will convert blue copper (cupric) salts to red cuprous compounds. The standard test for glucose is to add 8 drops of urine to 5 ml. of Benedict's qualitative sugar reagent and to boil vigorously for 3 minutes. A green, yellow or red precipitate indicates glucose.

In the reagent tablet method 5 drops of urine and 10 drops of water are placed in a tube and a tablet added. The reaction of the tablet with the water creates sufficient heat to cause boiling. The colour change is compared with a test chart. Both these tests give false positives with galactose.

In the reagent strip method, the strip is dipped into the urine. If glucose is present a colour is produced by enzyme action. This test is specific for glucose. Galactose does not give a positive result.

CLEARANCE

Clearance is a theoretical concept which is useful when considering kidney function. To explain it, let us consider the excretion of a substance in the urine, such as urea.

If we assume that 30 g. of urea are excreted per day, on the average, 20 mg. of urea will be excreted each minute.

The normal blood urea level is 30 mg. urea/100 ml. blood.

Thus 30 mg. of urea will be found in 100 ml. blood.

Therefore 20 mg. urea would be found in $\frac{20}{30} \times 100$ = 66 ml. blood.

Instead of considering that all the blood in the body loses 20 mg. of urea each minute, we could say that this is equivalent to 66 ml. blood being completely cleared of urea and the rest of the blood being unchanged. This figure of 66 ml. would then be the **urea clearance.**

(Urea clearance varies slightly with urine flow. It is about 54 ml./min. when the flow is 1 ml./min. and increases to a maximum clearance of 75 ml./min. during a diuresis when the flow is over 2 ml./min.)

Clearance is thus defined as the **blood volume that contains the quantity of a substance which is excreted in the urine per minute.**

This, at first sight, appears to be a rather pointless and confusing concept, but it enables the **kidney blood flow** and **glomerular filtration rate** to be determined in man.

The blood is completely cleared of some substances as it flows through the kidney. Following an injection, a substance such as **PAH** (para-aminohippuric acid) is both filtered in the glomerulus and secreted in the tubules, so that the blood, by the time it reaches the renal veins, has been completely cleared of PAH. The clearance of such a substance can be calculated, as above, from the blood concentration level and amount excreted in the urine per minute. The **clearance** will give the blood flow through the kidneys, since this volume of blood has actually passed through the kidneys and has been completely cleared of the substance.

The **PAH clearance** from blood is 1,200 ml./min. This determination of PAH clearances gives the **kidney blood flow.**

The clearance of PAH from the blood is only complete if the blood concentration is low. But a low concentration makes chemical analysis difficult. To overcome this difficulty, a derivative of PAH, *hippuran* (iodo-hippuric acid) containing radioactive iodine may be used. The concentration of this substance in blood and urine is estimated by the determination of its radioactivity. This principle allows extremely low doses to be employed.

The clearance of a substance, such as **inulin,** which is filtered, but is neither reabsorbed nor secreted in the tubules, will give the amount of plasma filtered per minute. The **inulin clearance** from plasma is 120 ml. per minute. This is the determination of the **glomerular filtration rate.**

Creatinine clearance is used clinically to give an approximate indication of the glomerular filtration rate. It is already present in the blood, and therefore has the advantage that, unlike inulin, it does not have to be given. In man, however, some tubular secretion of creatinine takes place, and too high a result is often obtained.

It follows that any substance with a clearance higher than that of inulin (e.g. penicillin), must be secreted by the kidney tubules into the urine in addition to being filtered by the glomeruli, whereas any substance with a lower clearance than inulin must be reabsorbed to some extent by the kidney tubules. It follows that since the clearance of urea is less than that of inulin, it is reabsorbed to a small extent even though it is a 'waste product'.

KIDNEY FAILURE

If the kidneys fail to produce an adequate volume of urine, the products of metabolism, and the surplus inorganic salts and water are retained in the body. The water and inorganic salt intake must then be limited to that lost by sweating and in the faeces.

A diet consisting of carbohydrate and fat is employed. The protein content must be reduced to the barest minimum. This is to prevent acidaemia from the accumulation of sulphuric acid in the body as a result of the breakdown of the surplus sulphur-containing amino acids methionine and cysteine [p. 64], and to prevent the accumulation of urea from the breakdown of surplus amino acids. The blood urea level is usually monitored and in renal failure it rises from a normal level of 30 milligrams urea per 100 ml. blood to 150 milligrams urea per 100 ml. blood or higher. Any changes in the acid-base balance, as indicated by the blood pH and bicarbonate level, should be immediately corrected.

There are no symptoms in kidney failure until the glomerular filtration rate has fallen from 120 ml. per minute to below 30 ml. per minute, that is, half of one kidney is still functioning. Below this dietary restrictions are needed. When the glomerular filtration rate has fallen from 120 ml. per minute to below 5 ml. per minute, control by diet alone is no longer effective, and an artificial kidney or a kidney transplant is needed to maintain life.

An artificial kidney uses the principle of dialysis. An artery (usually in the forearm) is cannulated and the patient's blood enters the artificial kidney. Here the blood is separated by a semi-permeable membrane from a dialysing fluid. The membrane allows the surplus inorganic salts, water, and metabolites to diffuse across into the dialysing fluid, but prevents the passage of blood cells or plasma proteins. The blood then re-

turns to the body via a forearm vein. The plastic cannulae are usually left in position and to prevent clotting when they are not in use, they are connected together to form an arteriovenous shunt.

The patient's own peritoneal cavity may be used as a semi-permeable membrane. The dialysing fluid is run into this cavity. This technique is known as **peritoneal dialysis.**

ORGAN TRANSPLANTATION

Kidney Transplants

The kidney was the first organ to be transplanted from a donor to a recipient. Its choice was no doubt influenced by the facts that there are a large number of patients in severe renal failure who could benefit from a kidney transplant, that everyone has two kidneys and yet is able to maintain normal urinary function with only one, and that a kidney transplantation is not technically difficult because the renal artery, renal vein and ureter are accessible without a complicated dissection.

A transplanted kidney is usually placed in the pelvis. The blood supply is obtained from the iliac blood vessels. The ureter is inserted directly into the bladder.

Antigen-antibody Reactions

The principal problem of tissue transplantation is one of rejection. The body is apparently able to recognize its own tissue, which it accepts, as different from foreign tissue which it rejects. The processes of rejection are complex. The rejection may be immediate or it may be delayed.

A foreign tissue acts as an **antigen** and stimulates the immunologically-active lymphocytes (B-lymphocytes) and their derivatives **plasma cells** to form a circulating **antibody.** This antibody is a globulin protein (**immunoglobulin**) which is found mainly in the gamma globulin fraction of the plasma proteins. Other, T-lymphocytes, which have been processed in the **thymus gland,** do not release their antibody, but destroy foreign cells by direct contact.

The antibody combines with the antigen, and this antigen-antibody reaction leads to the destruction of the antigen. This rejection takes many forms. A transplanted organ may be completely destroyed. A skin graft from a donor sloughs off after about three weeks due to this rejection process, whereas a skin graft from another part of the body takes and grows at the new site. In the case of foreign red blood cells (blood transfusion mismatch), the rejection is immediate and takes the form of agglutination of the foreign cells [p. 37].

If the antigen-antibody reaction is associated with the release of **histamine** in the skin, it will cause a skin rash and possibly wheal formation (triple response reaction, p. 21.) This is seen in angioneurotic oedema, and other allergic conditions. A severe form of histamine poisoning **anaphylactic shock** may follow the injection of a foreign protein to which the body has previously become sensitized. Care must be taken, for example, when giving antitetanus serum (ATS) to ensure that there will not be any untoward reaction on this account. A preliminary test dose is advisable. The effects of any histamine released can be reduced but not abolished by giving **antihistamines.**

The antigen need not be a protein. Even the metal nickel can act as an antigen and bring about an allergic response in the form of a skin rash under a nickel fastener. The patient may develop a sensitivity to penicillin.

Tissue Typing

When considering an organ transplant, the next most important grouping after the **ABO** red cell system appears to be the human leucocyte locus A (**HL-A**) system. **HL-A** antigens are found in most cells and tissues. Lymphocytes and platelets are particularly rich in these antigens.

Eleven **HL-A** antigens have so far been internationally recognized. Everyone has a total of four **HL-A** antigens, two from each series:

1st series 1, 2, 3, 9, 10, 11
2nd series 5, 7, 8, 12, 13

There are, as yet, no antigens 4 and 6.

A patient's **HL-A** typing may thus be expressed as: **2, 7 / 3,12**

Just as the determination of a patient's **ABO** grouping [p. 37] needs **ANTI-A** and **ANTI-B** antibodies (easy to obtain from the serum of **Group B** and **A** donors respectively), so the determination of the **HL-A** antigens needs a serum containing the **HL-A** antibodies.

The **HL-A** antibodies are obtained from:
1. Volunteers who have been given subcutaneous injections of lymphocytes carrying **HL-A** antigens.
2. Patients who have had a number of blood transfusions. Since the **HL-A** grouping is not considered when giving a blood transfusion, many of these transfusions will have been mismatches and cause the production of **HL-A** antibodies.
3. Women who have had many children and who have become immunized by the presence of foetal **HL-A** antigens from the father.
4. Another source is patients who have rejected kidney and other transplants which were **HL-A** incompatible.

The **HL-A** antibody destroys lymphocytes from a person of the same **HL-A** antigen group when they are incubated together. This fact is used to determine the **HL-A** groups of a patient. Alternatively platelets may be used.

However, poorly **HL-A** matched organs are not always rejected.

Suppression of Immunological Responses

Successful organ transplantation has only become possible with the development of means for suppressing the rejection processes. These are aimed principally at reducing the number of circulating lymphocytes.

Lymphocytes are formed by the lymphoid tissue and bone marrow. Lymphocyte formation, like that of the other blood cells is impaired by ionizing radiation such as X-rays and by certain cytotoxic chemicals [p. 35]. The lymphocytes can thus be reduced by irradiating the lymphoid tissue and by giving immuno-suppressive drugs. More recently antilymphocytic serum has been employed to destroy the lymphocytes.

As will be seen later [p. 104] the hormones of the adrenal cortex have an action in suppressing allergic responses, and high levels of cortisol have been employed as an additional measure to suppress rejection.

Unfortunately, the non-specific reduction in the circulating white cells, brought about by many of these procedures, increases the patient's liability to infection. Sterile conditions for the patient, with barrier nursing, are therefore essential following a transplantation. Although the level of immuno-suppressive drugs may be reduced after a time, a maintenance dose may be required for the rest of the patient's life.

TRANSPLANT OF ORGANS OTHER THAN THE KIDNEYS

The development of immuno-suppressive drugs has enabled the transplant of other organs to be undertaken. The heart beats because of the inherent properties of cardiac muscle. It will continue to beat when denervated. This means that in a heart transplant operation, no attempt need be made to suture the nerves running to the heart. Recent results indicate that provided the rejection problem can be overcome such a transplanted heart is able to maintain an adequate circulation. Animal experimentation leads to belief that the autonomic nervous system nerves will grow into and re innervate the transplanted heart after a few months. The transplanted heart will then be once again under the control of a cardiac centre in the medulla.

Other organ transplants include the **liver,** for the treatment of hepatic failure, and the **pancreas,** for the treatment of diabetes mellitus.

Some tissues of the body do not act as antigens and their transplantation presents no immunological problems. Such tissues include the cornea of the eye (corneal grafting), and bone.

There is no immunological reaction when an organ from an identical twin is transplanted.

It may be that in the future, it will be possible to undertake organ transplantation on the basis of 'organ groups' in the same way as blood is transfused on the basis of 'blood groups'. It is too soon to predict whether heart cells, kidney cells, liver cells, etc. will have their own grouping system.

14. TEMPERATURE REGULATION

Man is able to maintain his central body temperature, that is, the temperature of the brain, thoracic cavity and abdominal cavity, at a constant value which is independent of the temperature of the surrounding environment. In this way he differs from the cold-blooded animals whose body temperature varies with the surrounding environmental temperature.

Man has no fur on his skin, but he has subcutaneous fat which plays a similar role in conserving heat. To prevent an excessive loss of heat in a cold climate, he covers his skin with clothing. This clothing traps warm air close to the skin and creates a local external environment which man takes around with him. In this way man has become the only animal who can maintain his body temperature and thus survive in both the polar regions and the tropics.

MAINTENANCE OF A CONSTANT CENTRAL BODY TEMPERATURE

The constancy of the body temperature is achieved by continuously maintaining a balance between the heat which is gained and the heat which is lost [FIG. 111]. If

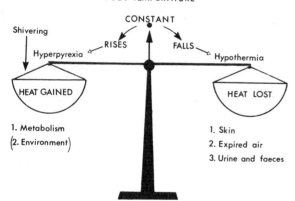

FIG. 111. The body temperature remains constant if the heat gained equals the heat lost. If the heat gained exceeds the heat lost the body temperature rises leading ultimately to hyperpyrexia. If the heat lost exceeds the heat gained the body temperature falls leading to hypothermia.

more heat is gained than is lost the body temperature will rise. If more heat is lost than is gained, the body temperature will fall.

Heat is gained by metabolism. At rest, the metabolic processes of the body produce about 70 Calories of heat per hour [p. 71]. If there were no heat loss the metabolic heat production would increase the body temperature by 1°C. (1·8°F.) every hour, and a fatal heat stroke would result in a very few hours.

Heat may also be **gained** from a hot environment, such as when sitting in the sun, or in front of a hot fire. A small amount of heat is gained by taking hot food.

In order to maintain a constant body temperature heat must be lost at the same rate as it is gained. Thus at rest 70 Calories of heat must be lost each hour. Heat is **lost** via the skin, in the expired air, and in the urine and faeces.

The principal **loss of heat** is via the **skin.** Heat is lost from the skin by radiation, conduction, convection and the evaporation of water.

1. Radiation

The skin radiates heat as infra-red rays which heat up any solid objects in their path. Thus when sitting in a room the radiant heat from a person's body is warming up the walls, floor and ceiling, and, of course, any other person who happens to be in the room.

The amount of heat lost by radiation depends on the colour of the clothing. Black objects lose (and gain) heat readily by radiation. White shiny objects do not. The silvery shiny suits worn by astronauts conserve heat when they leave the space capsule and are exposed to the intense cold of outer space. In the capsule, cooling fluid is pumped round inside the suit to prevent overheating when the suits are worn.

2. Conduction

Heat is also lost by conduction. Heat is conducted to the chair in which one is sitting or to any other object with which one is in contact. Heat is very rapidly lost in this manner to any object that is a good conductor of heat, such as a metal, but very little heat is lost through poor conductors of heat such as clothing and fabric covering to chairs.

3. Convection

Some heat is lost by convection. The air close to the skin is warmed, and this warm air rises and cold air comes in to take its place from the floor level. As a result of convection currents, the temperature of a room is often much higher at ceiling level than it is at floor level.

4. Evaporation of Water

The loss of heat by the evaporation of water from the skin is a very important way in which heat is lost in a warm environment. The sweat glands are controlled by the sympathetic nervous system [p. 119] and when active they pour sweat on to the surface of the skin. If this sweat evaporates then the skin will be cooled. If, on the other hand, this sweat runs off the body, then the cooling effect will be lost. The evaporation of 2 ml. of sweat on the skin will cause the loss of one Calorie of heat.

Sweating and the evaporation of sweat is a very efficient way of cooling the body when the surrounding air is dry. If the atmosphere, however, is humid, then the sweat does not evaporate and sweating no longer has a cooling effect.

Sweating is the only means available to the body of maintaining a normal body temperature when the environmental temperature is higher than 98·4°F. (37°C.). Under these conditions heat will be gained by the body, and not lost, by radiation, conduction and convection. Sweating will be needed to lose not only the heat produced by metabolism, but also to cancel out the heat gained by the other methods.

Provided that the air surrounding the body is dry, it is possible to maintain a normal body temperature at extremely high environmental temperatures, approaching that of boiling water. If the environmental humidity is very high, then a temperature of 80°F. (27°C.) may be very uncomfortable.

The expired air is fully saturated with water vapour, and heat is lost from the body by evaporating the water. The heat lost in this way is not very great under normal conditions, but may become quite high if the air breathed is excessively dry, or if the patient is over-breathing for any reason [see p. 78].

Urine and faeces leave the body at a temperature of 98·4°F. (37°C.) and it is possible to determine the body temperature of, for example, a child by measuring the temperature of the urine passed. The amount of heat lost by these two pathways is very small.

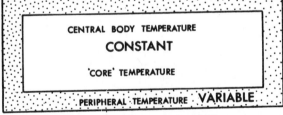

Skin temperature (variable)

ROOM TEMPERATURE (VARIABLE)

Fig. 112. The temperature in the brain, thorax and abdomen (central 'core' temperature) is constant. The skin temperature is not constant but is adjusted so that the heat lost via the skin maintains the body in thermal equilibrium.

TEMPERATURE REGULATION

The body temperature is regulated by the hypothalamus. If insufficient heat is being gained, then heat production can be increased by taking exercise. This will increase the metabolism. If no exercise is taken under these conditions then shivering occurs.

Shivering is the involuntary contraction and relaxation of the voluntary muscles, and it increases the heat production by increasing these muscles' metabolism.

Although the temperature of the central core of the body is maintained at a constant level, the temperature of the periphery and skin of the body is not a constant. Its temperature is varied in order to maintain the correct degree of heat loss [FIG. 112].

The heat lost from the skin is determined by the skin temperature. The higher the skin temperature the greater the heat lost, the lower the skin temperature the smaller the heat lost. The skin temperature depends on the skin blood flow and this, in turn, depends on the activity of the sympathetic nerves to the skin [p. 21]. The higher the sympathetic vasoconstrictor tone, the lower the skin blood flow, the lower the skin temperature and the lower the heat loss from the skin [FIG. 113].

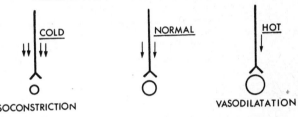

NERVE SUPPLY TO SKIN BLOOD VESSELS

VASOCONSTRICTION

Low skin blood flow
Low skin temperature
Low heat loss

VASODILATATION

High skin blood flow
High skin temperature
High heat loss

Fig. 113. The loss of heat from the skin depends on the skin blood flow. This is regulated by the heat regulating centre in the hypothalamus acting via the sympathetic nervous system to the skin. The intensity of the sympathetic activity is represented diagrammatically by the number of arrows alongside the nerve.

The lower the sympathetic vasoconstrictor tone, the higher the skin blood flow, the higher the skin temperature and the higher the heat loss.

When the body temperature is too high, such as following a hot bath, the skin blood vessels dilate so that the skin has an increased blood supply. The skin temperature rises and this increases the loss of heat by radiation, conduction and convection.

When the body temperature is low, the blood vessels of the skin are constricted, the skin becomes cold to the touch, and the loss of heat from the skin is reduced.

The skin temperature is thus not a constant like the central core temperature, but is adjusted in the interest of temperature regulation of the body as a whole [see p. 21]. The skin temperature may be as low as the environmental temperature in the extremities of the body such as the tip of the big toe; alternatively it may be approaching the central core temperature in areas such as the axilla and groin.

DETERMINING BODY TEMPERATURE

The standard way of determining the body temperature is to use the clinical thermometer. This is a 'maximum thermometer' which records the temperature either in degrees centigrade (Celsius) or in degrees Fahrenheit. A patient's temperature may be taken using either system. A conversion chart is given in FIGURE 114. The clinical thermometer has an arrow usually at 98·4°F. or 37°C. This represents the average body temperature. Incidentally, 37°C. = 98·6°F. and some Fahrenheit thermometers have the arrow at this reading.

Such a thermometer has its limitations. It is usually placed in the mouth under the tongue and therefore gives a reading of the mouth temperature. This will only be the same as the central core temperature if prior to taking the reading the mouth has been firmly closed so that breathing has only taken place through the nose, and provided that no hot or cold food has been placed

TABLE 5 shows typical variations in readings obtained when a '½ minute clinical thermometer' is used in an attempt to determine the central body temperature of an adult. It will be noted that the mouth temperature recorded after 1 minute was only 97·5°F. (36·4°C.), but that after replacing the thermometer in the mouth until a constant temperature was obtained, a final reading 1·3°F. (0·7°C.) higher, that is, 98·8°F. (37·1°C.) was recorded. As long as 20 minutes was needed in some cases to reach this more accurate reading.

The temperatures recorded in the axilla and groin were lower than that in the mouth.

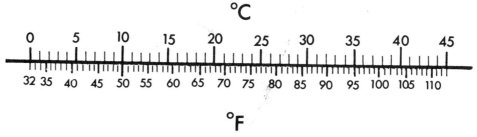

FIG. 114. A conversion chart between °C and °F. Temperatures may be recorded in either degrees Fahrenheit (°F) or degrees Centigrade (°C).

in the mouth. A cup of tea or coffee may be 160°F. (71°C.) and such a hot drink will greatly increase the mouth temperature for quite some time after it has been taken. Conversely, ice-cream at 32°F. (0°C.) or below will cool the mouth to a very low temperature. A mouth temperature is unreliable in exercise because of the increased respiration which involves mouth breathing. It is also unreliable if a patient has a head cold and is breathing through the mouth.

The clinical thermometer takes quite a long time to reach its final reading because of the large amount of heat required to be transferred to the mercury. For an accurate recording of temperature the thermometer should always be **replaced** until a **constant reading** is obtained. This is likely to take very much longer than the time 'half a minute' or 'three minutes' stated on the thermometer stem.

The wide range of readings that may be obtained can be readily confirmed by the reader. The only apparatus required is a clinical thermometer.

After a cup of hot tea (150°F. (65·6°C.)), the mouth temperature rose to 100·6°F. (38·1°C.). After 5 minutes breathing through the mouth, the mouth temperature fell to 96·8°F. (36·0°C.). These readings show errors of approximately 2°F. (1·1°C.) above and below the central body temperature.

The temperature recorded in the axilla or in the groin is often half to one degree lower than that recorded in the mouth. But it may give a more accurate indication than mouth temperature in many of the cases referred to above.

The rectal temperature is often half to one degree higher than that recorded in the mouth. But it may be an inaccurate indication of a change in body temperature if the thermometer is impacted in faeces.

Under normal conditions an approximate indication of the central body temperature can be obtained using a thermistor type of thermometer, in the external acoustic meatus of the ear, attached to the skin under the arms, or in the umbilicus.

Diurnal Variations in Body Temperature

Although the body temperature is approximately constant at 98·4°F. (36·9°C.), it fluctuates throughout the 24 hours reaching a maximum usually about 6 p.m. and a minimum round about 4 a.m. when a person is active during the day. These are termed diurnal variations. After exercise the body temperature may be in excess of 100°F. (37·8°C.), and a temperature even higher than this may be the result of a hot bath. In women the temperature may be one degree higher after ovulation in the latter part of the menstrual cycle than

TABLE 5

Temperatures recorded with a '½ minute' clinical thermometer at a room temperature of 70°F. (21°C.) (mean of observations on 15 subjects)

MOUTH	AXILLA	GROIN	MOUTH AFTER HOT DRINK	MOUTH AFTER MOUTH BREATHING
97·5°F. 98·8°F. 36·4°C. 37·1°C. (1 min.) (20 min.)	98·0°F. 36·7°C.	98·0°F. 36·7°C.	100·6°F. 38·1°C.	96·8°F. 36·0°C.

it is before ovulation. Daily recordings of early morning temperature are used to determine the time of ovulation, when studying cases of infertility.

Hypothermia

Young babies and old people, if exposed to severe cold, may go into a state of hypothermia, and the body temperature may fall below the lowest reading on the clinical thermometer, that is, below 95°F. (35°C.). In such a case a special hypothermia thermometer reading down to 80°F. (26·7°C.) is needed in order to determine the body temperature.

A new-born baby has brown body fat which is metabolized if the baby becomes cold, and safeguards the baby from hypothermia. This brown body fat, however, is used up in the first few days of life, and from then onwards it is important to ensure that the baby is not subjected to excessive cold.

A reduced body temperature leads to coma, and once this has occurred, exercise and muscle contractions will cease so that the metabolism is further reduced. As a result the body temperature falls more rapidly.

A hot bath is an effective way of raising the body temperature from hypothermic to the normal levels.

Hypothermia reduces the rate of metabolism of the body and reduces the blood flow requirements of tissues. This applies particularly to the brain. Under controlled hypothermia the brain blood supply may be interrupted for a few minutes without causing the permanent damage to brain cells that follows cerebral ischaemia at a normal body temperature. Using hypothermia it has been possible to stop the heart for a short time to enable cardiac surgery to be carried out. This technique is tending to be replaced, however, by the use of mechanical pumps which by-pass the heart and enable the cerebral blood flow to be maintained.

When a subject is undergoing an operation involving the intentional lowering of body temperature to a state of **hypothermia,** it is essential to record the body temperature continuously. Thermocouples and thermistor thermometers may be employed since these give an almost instantaneous reading of the temperature. But a number of these, recording the temperature in different parts of the body, are required for an accurate assessment to be made. If only a single thermometer is used, the best guide to the central body temperature is probably obtained with an oesophageal thermometer.

Hyperpyrexia

The toxins from many infections act on the temperature regulating centre and cause it to be set at a higher level. As a result the body temperature rises, and a fever develops.

During the rise in body temperature the heat lost is less than the heat gained. The heat lost is reduced by a reduction in the skin blood flow brought about by an increase in sympathetic vasoconstrictor tone [p. 21]. In addition the heat gained may be increased by shivering. The patient feels cold due to the fall in skin temperature associated with the reduction in skin blood flow, and this combined with the shivering constitutes a **rigor.** Once the new increased temperature has been reached, a new equilibrium is set up with the heat gained equal to the heat lost. The skin vessels dilate and the patient then feels warm.

Drugs which bring about a fall in body temperature under these conditions of hyperpyrexia are termed **antipyretics.** They include acetylsalicylic acid (aspirin) and paracetamol.

Should the temperature in a fever continue to rise, cooling of the skin by the evaporation of water may be necessary to prevent the development of a fatal hyperpyrexia. Sponging with warm water is often employed, as sponging with cold water causes skin vasoconstriction and is less comfortable for the patient. The evaporation of water can be speeded by increasing the rate of airflow over the patient using an electric fan.

When the temperature starts to fall to normal after a fever, the body often loses heat rapidly. This is achieved by dilatation of the skin blood vessels. The skin becomes warm, and in addition sweating may occur. The patient feels very hot due to the vasodilatation which raises the skin temperature and stimulates heat receptors in the skin [p. 126].

Thus a patient may feel cold when the temperature is going up, and hot and sweaty when the temperature is coming down.

15. THE DIGESTIVE TRACT AND DISORDERS OF DIGESTIVE FUNCTION

The food passes from the mouth down the oesophagus to the stomach. From the stomach it proceeds to the duodenum, jejunum and ileum (small intestine). The unabsorbed residue passes via the caecum, ascending, transverse, descending, and sigmoid colons (large intestine) to the rectum and anal canal.

MOUTH—SALIVA

Saliva is produced by three salivary glands: the **parotid,** the **submandibular** and the **sublingual** glands. Saliva is secreted in response to the sight, thought, taste or smell of food. Salivation is brought about by the parasympathetic nervous system [CHAPTER 19, p. 117]. The parasympathetic fibres in the ninth cranial nerve supply the **parotid gland.** The parasympathetic fibres, in the seventh cranial nerves, supply the **submandibular** and **sublingual salivary glands.** These nerves are secretomotor to the salivary glands, that is, they bring about an increased production of saliva. As has already been seen [p. 21] these nerves also bring about vasodilatation of the blood vessels leading to the salivary glands so as to provide a larger blood flow during salivation.

Parasympathetic activity is blocked by atropine and hyoscine (scopolamine) and either of these drugs will give a dry mouth.

Saliva has many **functions.** It acts as a lubricant, and enables a **bolus** to be formed. It contains **salivary amylase,** which is also known as **ptyalin.** It has a cleansing action. A patient with a fever has a reduced production of saliva, and as a result often has a coated tongue. It makes speech possible; it is not possible to talk if the mouth is dry. It enables the **taste buds** to respond to sweet, salt, acid and bitter substances: the sensation is only possible if the substance can dissolve in water. But, as will be seen later [CHAPTER 23], many tastes are, in fact, smells due to the stimulation of the olfactory receptors in the nose. When one has a cold in the head this sense of smell is impaired, but the taste buds are unaffected. The alteration in the taste of a food under these conditions is due to the loss of the olfactory component.

The saliva is fully saturated with calcium and this prevents decalcification of the teeth. Any acid placed in the mouth will tend to dissolve the teeth. If a person has to be given, for example, hydrochloric acid by mouth, it must be taken through a straw and not allowed to touch the teeth. The mouth must be washed out thoroughly with water afterwards. The formation of acids in the mouth, as a result of bacterial action, probably plays an important role in dental decay.

OESOPHAGUS

The food is formed into a bolus in the mouth, and is then swallowed. It passes down the oesophagus by means of a peristaltic movement brought about by the vagus nerve. As a result of this peristalsis, it is possible to swallow both liquids and solids whilst lying down, and even if standing on one's head!

The oesophagus is lined with stratified squamous epithelium, that is, like skin, and the nerve endings in the oesophagus are well localized in the brain [p. 128]. The passage of a very hot drink down the oesophagus can be followed. A patient with an oesophageal obstruction is often able to tell quite accurately where the obstruction actually is.

Once the food reaches the stomach, and passes into the intestine, localization is no longer possible and it is not possible to tell where the food has reached in this part of the digestive tract.

The oesophagus, unlike the intestines, has no outer peritoneal coat surrounding it. This makes a watertight anastomosis more difficult in oesophageal surgery.

STOMACH—GASTRIC JUICE

The food enters the stomach via the **cardiac sphincter** [FIG. 115]. This is not a sphincter that can be demonstrated anatomically but this part of the digestive tract acts as a physiological sphincter which prevents the regurgitation of food from the stomach up the oesophagus. Should this occur, then the gastric contents

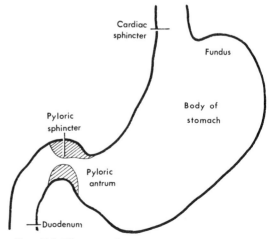

FIG. 115. The stomach. The 'physiological' cardiac sphincter prevents the regurgitation of gastric contents up the oesophagus. The 'anatomical' pyloric sphincter allows the stomach's content to be passed on gradually to the duodenum.

will irritate the oesophagus and lead to **oesophagitis.** Such regurgitation may occur with a **diaphragmatic (hiatus) hernia.**

Achalasia of the cardia is the name given to a condition where the cardiac sphincter does not open completely, and the swallowed food is obstructed at this point. One treatment of this condition is for the patient to swallow a mercury bougie before each meal. The weight of the mercury opens the cardiac sphincter and the food is then able to enter the stomach.

The **stomach** acts as a **storage organ** so that a large amount of food can be eaten at one meal. Every few minutes a small quantity of this food is passed through the **pyloric sphincter** into the duodenum until the stomach is empty. This process may take 2–4 hours, during which time the small intestine is provided with a continuous supply of nutrients for digestion and reabsorption.

A fatty meal causes the release of *enterogastrone* which inhibits the emptying of the stomach.

The emptying of the stomach may be delayed by fear. It is delayed when the pyloric sphincter is narrowed and does not open completely. This condition is known as **pyloric stenosis.** Such a condition may be congenital (present at birth). One form of treatment (Ramstedt's operation) consists of making a longitudinal incision in the pyloric sphincter muscle but leaving the underlying mucosa intact (like cutting a ring to remove it from the finger).

The **stomach** produces **pepsin** and **hydrochloric acid,** for the digestion of protein. In some animals (not man) it produces **rennin** (rennet) which curdles milk.

Gastric secretion is under the combined control of the **vagus nerve,** and the hormone **gastrin.** The vagus starts the flow of gastric juice at the sight and thought of food and when food enters the mouth. Gastrin is released from the pyloric region of the stomach into the blood stream when food reaches this part of the stomach. The gastrin circulates round in the blood and returns to the secreting cells of the stomach augmenting the activity of the vagus in the production of gastric juice.

Gastrin is a peptide containing 17 amino acids. Synthetic compounds having only the same terminal four or five amino acids have a similar action. **Pentagastrin** is such a compound.

Peptic Ulcer

Since pepsin is a protein-digesting enzyme, it may digest the stomach wall or the first part of the duodenum. When this occurs, the condition is termed a **peptic ulcer.**

If the ulcer is in the stomach it is referred to as a **gastric ulcer.** If it is in the duodenum it is termed a **duodenal ulcer.**

The complications of a peptic ulcer are perforation, haemorrhage and obstruction.

The acid output of the stomach is studied using a **fractional test meal.** The stomach is aspirated every 15 minutes using a stomach tube following the stimulation of gastric secretion by either **histamine** or **pentagastrin.** Duodenal ulcer patients frequently have a maximum acid concentration that is higher than normal. (They also have a higher pepsin and intrinsic factor level.) Gastric ulcer patients are usually no different from normal.

The presence of an ulcer cavity may be shown up on X-ray after the ingestion of insoluble barium sulphate (**barium meal**). It may also be seen using an optical instrument passed down the oesophagus known as a **gastroscope.** Modern gastroscopes employ fibre optics to which a photographic or television camera can be attached.

The treatment of a peptic ulcer involves the reduction in the production of gastric juice by the stomach. This may be brought about in a number of ways. The vagus nerve may be cut (**vagotomy**) to reduce the vagal activity. The pyloric region may be excised to reduce the production of gastrin. A partial gastrectomy may be carried out to reduce the amount of gastric juices produced. Alternatively the hydrochloric acid may be neutralized by giving protein such as milk, or by means of alkalis such as aluminium hydroxide, or magnesium trisilicate. A more powerful alkali such as sodium bicarbonate is not so effective as it often brings about an acid rebound, that is, an increased production of acid some time later.

Vagotomy

The vagus nerve is cut as it enters the abdominal cavity. The fibres to the heart and other thoracic organs are thus still intact. The vagus has a wide distribution in the abdomen [p. 118] and a complete vagotomy will denervate the pancreas, gall-bladder and intestines as well as the stomach. This may lead to steatorrhoea and post-vagotomy diarrhoea. In a **selective vagotomy** only the branches to the stomach are cut.

Vagotomy, in addition to reducing the acid-pepsin production of the stomach, prevents the normal gastric emptying. As a result, either the stomach has to be joined to the jejunum by a **gastro-enterostomy** to drain the stomach, or alternatively the pylorus has to be permanently opened by a **pyloroplasty.** In this operation the pyloric sphincter is replaced by a large aperture between the stomach and the duodenum.

Partial Gastrectomy

When a partial gastrectomy is carried out for the treatment of a gastric or duodenal ulcer, two-thirds to three-quarters of the stomach is removed. This includes most of the acid-pepsin producing part of the stomach plus the pyloric antrum. The duodenum is closed off, and the remainder of the stomach is joined to a loop of jejunum in front of, or behind, the colon. This procedure has nutritional and metabolic consequences.

The storage function of the stomach will be lost and frequent small meals will be required until the patient has achieved an alternative storage site by distending part of the small intestine.

The loss of the **intrinsic factor** will lead to **pernicious anaemia** when the liver's reserve of B_{12} has been used up.

There will be inadequate mixing of bile and pancreatic secretion with the food, and the normal gastro-intestinal hormonal (humoral agent) release (gastrin, secretin, pancreozymin—cholecystokinin) will be upset.

A distressing sequela is termed the **dumping syndrome.** The sudden entry of hyperosmotic food into the jejunum sucks water from the plasma into the intestine by osmosis, thus causing intestinal distension and plasma depletion. In addition to abdominal distension the patient complains of giddiness, faintness and palpitations.

PANCREAS—PANCREATIC JUICE

The pancreas [FIG. 116] is both an exocrine gland, and an endocrine gland. Its exocrine function is to produce pancreatic juice, which enters the duodenum together with the bile. The endocrine function is the production of the hormones **insulin** and **glucagon.** A failure to produce insulin in sufficient amounts leads to diabetes mellitus [p. 70].

Pancreatic juice contains **trypsinogen** and **chymotrypsinogen** which are precursors of the protein-splitting enzymes **trypsin** and **chymotrypsin.** The trypsinogen is activated by **enterokinase** from the duodenal mucosa to form trypsin. The trypsin activates the chymotrypsinogen to form chymotrypsin.

The pancreatic juice also contains amylase and maltase for the digestion of carbohydrates, and lipase for the digestion of fat.

Pancreatic secretion is under the control of the vagus, and the two humoral agents **secretin** and **pancreozymin** which are released when food enters the duodenum.

A tumour of the cells of the islet of Langerhans of the pancreas which produce a gastrin-like substance gives rise to the **Zollinger-Ellison syndrome.**

LIVER—BILE

The bile which is made in the liver enters the duodenum via the bile-duct [FIG. 116]. It contains the **bile salts** for

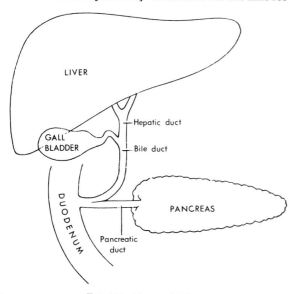

FIG. 116. Liver and biliary tract.

the emulsification of fat. The bile salts are the sodium salts of compounds of glycine or taurine (amino acids) and cholic acid (a bile acid). They are thus termed glycocholates and taurocholates. As has been seen these bile salts break the fat down into minute fat droplets which will mix with water. The bile salts are reabsorbed in the ileum by a special transport mechanism and returned to the liver via the portal vein. They are then resecreted down the bile-duct. The bile salts recirculate round in this fashion about six times a day.

The bile is also a route of excretion for the breakdown products of red blood cells [p. 35] in the form of bilirubin and biliverdin. These are known as the **bile pigments.**

The bile also contains **cholesterol.**

The bile is concentrated and stored in the gall-bladder. When fatty foods enter the duodenum, the humoral agent **cholecystokinin** (which appears to be identical to pancreozymin) is released. It circulates round to the gall-bladder and causes it to contract. The contents of the gall-bladder then pass down the bile-duct and enter the duodenum. Alternatively, the gall-bladder will contract as a result of vagal stimulation.

The constituents of bile may be precipitated in the gall-bladder in the form of **gall-stones.** Thus if cholesterol is deposited then a cholesterol stone is formed. This is particularly likely to occur following inflammation of the gall-bladder (**cholecystitis**). A gall-stone in the gall-bladder is comparitively harmless. It is only when the gall-stone passes along the cystic duct and becomes impacted in the common bile-duct that jaundice will occur [FIG. 61, p. 36]. The jaundice is due to the body's inability to excrete the bilirubin, due to the obstruction of the bile-duct. This obstruction is removed by surgery. The gall-stone may pass spontaneously down the bile-duct into the duodenum, in which case the jaundice will disappear.

This is discussed further in the next chapter.

SMALL INTESTINE

The secretion of digestive juices in the small intestine is due to the presence of food in that part of the digestive tract.

The small intestine produces the enzymes **maltase, sucrase** and **lactase** for the conversion of disaccharides into monosaccharides, and the **collection of peptidases,** erepsin, to complete the breakdown of proteins to amino acids. These enzymes are found principally in the cells lining the intestine, and the breakdown is probably completed after absorption into the cells.

The small intestine has a nerve supply from both the sympathetic and the parasympathetic nervous systems, but these nerves play no part in the release of digestive enzymes. However, they play an important role in regulating the intestinal motility. An increase in parasympathetic activity speeds up the movements of the intestine, whereas an increase in sympathetic activity reduces these movements. These movements consist of pendular contractions of the intestinal wall, followed occasionally by a peristaltic movement which moves the food along. Should the movement cease the condition is termed **paralytic ileus.** It may be treated by augmenting the parasympathetic nervous system activity using a drug such as **carbachol** (an acetylcholine derivative).

Intestinal Absorption

The absorption of food takes place mainly in the small intestine. This process is partly one of diffusion from a region of high concentration in the lumen of the intestine to a low concentration in the intestinal cells and blood, and partly one of **active transport** by the cells. The substance, such as a sugar, is taken in on one side of the cell next to the lumen, transported across the cell and secreted into the blood capillaries on the other side. Such a transport system uses chemical compounds as carriers.

There is a limit to the quantity of the sugar that these carriers can transport in a given time [see kidney tubule cells, p. 85]. The system requires energy which is obtained from metabolism in the intestinal cells.

Vitamin B_{12} is absorbed by a special transport mechanism with the aid of the intrinsic factor [p. 35] in the region of the terminal ileum. With a large intake of vitamin B_{12}, such as by ingesting liver daily, absorption from the rest of the small intestine may be sufficient without the need for the intrinsic factor and this special transport mechanism.

LARGE INTESTINE

The food passes through the **ileocaecal valve** into the large intestine. Each day about 500 ml. of intestinal content enter the large intestine, and during the passage through the large intestine the volume is reduced by the absorption of water to about 150 ml.

The large intestine is not essential to life and the digestive tract may be terminated at the end of the small intestine (**ileostomy**).

The bacteria which live in the large intestine provide the body with a source of **vitamins,** particularly those of the B group, which are absorbed, and reduce the need for an intake of these vitamins [p. 73]. Following the administration of a **broad-spectrum antibiotic** these bacteria will be destroyed, and a **vitamin deficiency** may result unless the vitamin intake is supplemented.

The large intestine acts as a storage organ for the accumulation of food residue.

The entry of the faeces into the **rectum** brings about a desire for defaecation. In addition, entry of food into the stomach causes contraction of the colon and a desire for defaecation (**gastrocolic reflex**).

Defaecation is brought about by the action of the parasympathetic nervous system, which contracts this part of the large intestine, and relaxes the **internal anal sphincter.** In a baby defaecation is an autonomic reflex, but, after a year or so of life, the **external anal sphincter** comes under voluntary control and now defaecation does not occur unless this external sphincter is also relaxed. This external sphincter is controlled by higher centres. If a person has a brain injury or spinal cord damage this higher centre control may be lost. Defaecation will then revert back to being an autonomic reflex which occurs whenever faeces enter the rectum.

16. THE LIVER

The blood returning from the intestinal tract reaches the liver via the portal vein. In addition the liver has a blood supply from the abdominal aorta via the hepatic artery. The blood from these two vessels passes through sinusoids lined by liver cells as it flows to the hepatic vein.

The liver is arranged in lobules. In each lobule the liver cells are arranged like spokes of a wheel around a central vein which is a tributary of the hepatic vein leading to the inferior vena cava and the right side of the heart.

The blood as it passes through the sinusoids is in direct contact with the liver cells since the capillary wall is absent. This enables substances to be extracted readily from the blood by the liver cells. Certain materials such

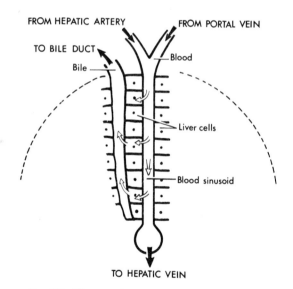

FIG. 117. Diagram of a lobule of the liver. Liver cells are arranged radially around the central hepatic vein.

as the bile pigments pass through the cells and are excreted into the bile-duct on the other side of the cells [FIG. 117].

The main blood flow is from the portal vein. It appears that the hepatic artery vessels are normally closed off by sphincters and that they open intermittently for a brief period of time sending a spurt of blood through the sinusoids to the hepatic vein.

FUNCTIONS OF THE LIVER

The liver is essential to life. It is the chemical factory of the body and has numerous functions. They may be conveniently grouped under the headings of:

1. Blood
2. Food
3. Foreign substances.

Blood

The liver plays an important part in the formation and destruction of red blood cells. In foetal life it is a site of formation of red blood cells.

It stores vitamin B_{12} which passes when required to the bone marrow for the normal maturation of red blood cells.

It removes bilirubin, formed from broken-down red blood cells, from the blood and excretes this bilirubin down the bile-duct into the duodenum. If this excretory pathway fails, the bilirubin accumulates in the blood and jaundice develops [p. 36].

The liver manufactures plasma proteins, particularly the albumin fraction.

It forms the blood clotting factors prothrombin and fibrinogen.

Food

1. **Carbohydrate.** The liver plays an important part in maintaining the blood glucose level. It does so by converting excess carbohydrate to liver glycogen and also by forming glycogen from surplus protein and fat. This liver glycogen is used to maintain the normal blood glucose level as glucose is used up. Failure or removal of the liver leads to a fatal fall in blood glucose level (hypoglycaemia).

The liver converts galactose, which is derived from lactose the sugar in milk, into glucose. The lactating mammary gland reverses the process and converts blood glucose into galactose for the formation of lactose.

2. **Protein.** The liver forms urea from the ammonia which results from the deamination of the surplus amino acids. This enables the carbon and hydrogen part of these amino acids to be used for heat and energy. The urea is excreted via the kidneys in the urine. In liver failure the blood urea level falls, but the blood ammonia level rises.

3. **Fat.** The bile salts produced by the liver play an important part, in conjunction with pancreatic lipase, in the digestion and absorption of fat [p. 69]. The liver forms ketone bodies when fat is being metabolized in the absence of sufficient carbohydrate metabolism [p. 69]. It stores the fat-soluble vitamins (A and D).

Foreign Substances

The liver plays an important part in modifying drugs so that they can be excreted readily by the kidneys. Some substances, such as short-acting barbiturates, are completely destroyed by the liver, whilst other substances are made more soluble by conjugation. They are joined, for example, to glycine, glucuronic acid (a glucose derivative), sulphuric acid or acetic acid.

The duration of action of hypnotic drugs may be unduly prolonged if this 'detoxication' is retarded in liver disease.

Bilirubin Excretion

200 mg. bilirubin are made each day from broken-down red cells [p. 35]. Bilirubin is insoluble in water. It is kept in solution in the plasma by being combined with plasma albumin in its passage to the liver. This *pre-hepatic* bilirubin-protein combination is not filtered by the glomeruli of the kidney. Hence no bilirubin appears in the urine in haemolytic anaemia even though there is a marked increase in red cell breakdown and circulating bilirubin in this condition.

In the liver, the albumin is removed and replaced by glucuronic acid (an acid made from glucose). The water-soluble bilirubin-glucuronide passes down the bile-duct to the duodenum as *bile pigments* for excretion in the faeces. In obstructive jaundice (due to gall-stones, etc.) this form of post-hepatic bilirubin is re-absorbed into the blood where it circulates as plasma bilirubin-glucuronide.

This post-hepatic bilirubin-glucuronide gives an immediate red colour when diazotized with sulphanilic acid and is referred to as 'direct reacting bilirubin'. The test is termed the *Direct Van Den Bergh Reaction*.

The pre-hepatic bilirubin-protein combination only gives the colour after alcohol has been added to remove the albumin. Pre-hepatic bilirubin is termed 'indirect reacting bilirubin'. The test is termed the *Indirect Van Den Bergh Reaction*.

Jaundice

A failure to excrete bilirubin gives rise to jaundice. The cause may be:

1. Before the liver—**pre-hepatic jaundice.** This form of jaundice is due to an excessive breakdown of red cells, that is, haemolysis [p. 36]. An example is the physiological jaundice which occurs in a baby shortly after birth due to rapid red cell breakdown as the red cell count falls from 7 million to 5 million red cells per cubic millimetre. A more severe form occurs with a Rhesus mismatch, and later in life with haemolytic anaemia.

2. In the liver itself—**hepatocellular jaundice.** This form of jaundice includes viral jaundice and jaundice induced by drugs and poisons.

Viral jaundice appears to be caused by two viruses, one with a short incubation period (virus A) which gives rise to infectious hepatitis, and another with a long incubation period (virus B) which is transmitted via blood transfusions and gives rise to serum hepatitis. Virus B can sometimes be detected in a carrier by the presence of a hepatitis-associated antigen in the blood known as the Australia antigen. Presumably such a person should not be allowed to be a blood donor. Outbreaks of viral hepatitis are a potential hazard to patients and staff in renal haemodialysis units [p. 86].

Many poisons including pesticides, solvents such as carbon tetrachloride and dry cleaning fluids damage the liver and may cause jaundice. In susceptible patients drugs as widely varied as long-acting sulphonamides, mono-amine oxidase inhibitors, muscle relaxants, halothane and anabolic steroids may have the same effect and induce jaundice.

3. After the liver—**post-hepatic** or **obstructive jaundice.** This form of jaundice usually involves the blockage of the common bile-duct due to an impacted gall-stone [p. 36] or carcinoma of the head of the pancreas.

Liver Failure (Hepatic Insufficiency)

The liver is indispensable for life. If it fails or is removed, the blood glucose level falls. This will lead to a fatal hypoglycaemia unless the blood glucose level is maintained by an intravenous transfusion of glucose. The blood urea falls since the liver is the principal site of urea formation, but deamination of surplus amino acids continues and the blood ammonia rises [p. 66]. Bilirubin is not excreted and jaundice develops. Any drugs given will not be detoxicated and may have an excessively prolonged action.

Portal Hypertension

A raised pressure in excess of 10 mm. Hg in the portal vein is termed portal hypertension. It results from cirrhosis of the liver which causes obstruction of the intralobular tributaries of the hepatic veins. Occasionally the obstruction to the venous return via the liver to the right side of the heart may be due to a thrombosis in the portal vein itself or to its replacement by a cavernous mass of small venous channels.

A high portal venous pressure leads to ascites, hypersplenism and liver failure with ammonia intoxication and jaundice. It also tends to open anastomoses between the portal and systemic circulations leading to varices at the lower end of the oesophagus and upper part of the stomach. Bleeding oesophageal varices may cause haematemesis (vomiting blood). Rarely anastomoses around the umbilicus may give rise to a *caput medusae* and anastomoses around the rectum to haemorrhoids.

One form of treatment is a portacaval shunt, that is, the surgical creation of an anastomosis between the portal vein and the inferior vena cava to drop the pressure in the portal vein. But the liver then has a greatly reduced blood supply and its ability to remove ammonia from the circulation will be further reduced. Ammonia intoxication may lead to coma (porto-systemic encephalopathy).

Liver Damage

As has been seen [p. 64] the liver is able to change one amino acid into another. It is able, for example, to remove the amino group from the amino acid **glutamic acid,** and to attach it to pyruvic acid (which has no amino group) making a new amino acid **alanine.** The enzyme in the liver cells which carries out this conversion is known as **glutamic-pyruvic transaminase.** This enzyme is liberated into the blood when liver cells are damaged. The level of the **serum glutamic-pyruvic transaminase (SGPT)** is one of the determinations carried out in suspected liver disease.

Since the same enzyme will also reconvert **alanine** to **glutamic acid,** it is also called **alanine transaminase.**

Another enzyme known as **aspartate transaminase** or **glutamic-oxaloacetic transaminase** converts the amino acid **aspartic acid** into **glutamic acid.** The level of **serum glutamic-oxaloacetic transaminase (SGOT)** rises in liver disease and following a coronary thrombosis (since it is also released from damaged cardiac muscle).

Liver Disease and Plasma Protein Formation

The electrophoretic pattern of plasma proteins shows five bands: albumin, α_1 globulin, α_2 globulin, β globulin and γ globulin. In liver disease the formation of plasma proteins is reduced. The production of albumin is impaired to a greater extent than that of globulin (which is also made outside the liver) so that the level of globulin may exceed that of albumin in the plasma. These changes are shown clearly on the electrophoretic pattern.

The clotting factors fibrinogen and prothrombin are also affected and a deficiency may reduce the ability of the blood to clot.

The reduction in albumin may lead to oedema [p. 43].

17. HORMONES

CONTROL SYSTEMS

The activity of different organs of the body is controlled in two ways. The first system of control is by means of **hormones**. These are chemical substances, produced by endocrine glands, which circulate in the blood and modify the activity of distant organs. The second system of control is by means of nerve impulses transmitted along **nerves**. Some organs are under the simultaneous influence of both nerves and hormones. The heart, for example, is under the control of nerve impulses passing along the vagus and sympathetic nerves as well as hormones, noradrenaline and adrenaline, which are released as hormones by the adrenal medulla.

Introduction of the Term 'Hormone'

In 1902, Bayliss and Starling whilst working on the control of digestive secretions found that an extract from the intestinal mucosa contained a substance which when injected intravenously brought about a secretion of pancreatic juice. They went on to show that this substance which they termed **secretin** was released into the blood when food entered the small intestine. Following discussions with colleagues at Cambridge, the word **hormone** (from the Greek *hormaō* to excite) was coined for such substances which circulate in the blood and bring about an effect on a distant organ.

The word has tended more recently to be restricted to the substances produced by the endocrine glands, although the digestive tract humoral agents (secretin, gastrin, pancreozymin, etc., p. 92) are strictly speaking, hormones.

ENDOCRINE GLANDS

The secreting glands in the body are divided into two types. Those which pass their secretion along ducts, such as sweat glands and salivary glands, are termed **exocrine glands.** Those which are ductless and pass their secretion into the blood are termed **endocrine glands.**

The endocrine glands are:

Pituitary
Thyroid
Four parathyroids
Two adrenals (suprarenals)
Two gonads (ovaries in the female, testes in the male)
Placenta in pregnancy
Pancreas—both an endocrine gland producing insulin and an exocrine gland producing pancreatic juice.

THE PITUITARY GLAND [HYPOPHYSIS]

The pituitary gland lies in a bony cavity at the base of the skull known as the pituitary fossa. The gland is suspended from the hypothalamus by the pituitary stalk.

The pituitary gland consists of two parts: the posterior pituitary or neurohypophysis and the anterior pituitary or adenohypophysis. The two parts of the pituitary gland develop separately, and later fuse. The posterior pituitary is a downgrowth from the brain whereas the anterior pituitary starts as an upgrowth from the mouth and nasal cavity. The two parts which have entirely different functions will be considered separately.

THE POSTERIOR PITUITARY GLAND

The posterior pituitary is composed mainly of nervous tissue. These nerves have their cells of origin in the supra-optic nucleus and surrounding nuclei of the hypothalamus. The nerve fibres run down the pituitary stalk to terminate in the posterior pituitary itself [FIG. 118].

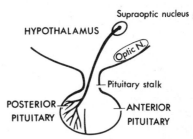

FIG. 118. The posterior pituitary hormones are made and released by nerves which run down the pituitary stalk. These nerves originate in the hypothalamus and constitute the hypothalamo-hypophyseal tract. This release of hormones by nerves is an example of a neurosecretory system.

The chiasma of the optic nerve lies immediately in front of the pituitary gland, and as a result a tumour of this gland may lead to loss of the visual field or complete blindness due to pressure on the optic nerve [p. 139].

The posterior pituitary gland produces two hormones:

1. Antidiuretic hormone (ADH, vasopressin).
2. Oxytocic hormone (oxytocin).

Antidiuretic Hormone (Vasopressin)

Antidiuretic hormone ADH is a small peptide made up of nine amino acids. Its function is to regulate the reabsorption of water by the kidney tubules [p. 79].

Injection of large amounts of ADH causes hypertension, due to its vasoconstriction action on blood vessels. It is thus a vasopressor substance, and for this reason it is also termed 'vasopressin'. There is, however, no evidence that the normal levels of circulating ADH have any effect on blood pressure, whereas they do have a marked effect on the volume of urine passed.

The release of ADH is under the control of the hypothalamus. The hormone itself is made by the nerves which run from the hypothalamus to the posterior pituitary, and is released in response to nerve impulses. This constitutes a control system that is a combination of both the nervous and hormonal systems. It is termed a **neurosecretory system.**

The release of ADH is regulated by the amount of water in the blood. The smaller the amount of water in the blood, the greater will be the osmotic pressure exerted by its constituents. Osmoreceptors are found in the vicinity of the arteries supplying the hypothalamus and these receptors are sensitive to osmotic pressure. They sample the osmotic pressure of the blood and transmit nerve impulses to the supra-optic nucleus of the hypothalamus. The activity of the osmoreceptor nerves regulates the activity of the nerve fibres running down the hypothalamus to the posterior pituitary gland (hypothalamo-hypophyseal tract).

If the body is short of water more ADH is released and this increases the reabsorption of water by the kidney tubules so that less urine is produced. If there is a surplus of water in the body then the ADH level falls, less water is reabsorbed by the tubules and more urine is produced [see water balance, p. 78].

If the posterior pituitary fails to produce ADH then **diabetes insipidus** develops. The subject is excessively thirsty and large quantities of urine are produced [p. 79].

Following the removal of the pituitary gland (hypophysectomy) a patient will need a replacement therapy for all the anterior pituitary hormones, but may not develop diabetes insipidus. Presumably in such a case ADH is still being produced by the hypothalamus and the pituitary stalk.

Oxytocic Hormone (Oxytocin)

The oxytocic hormone differs from ADH in only two of its nine amino acids. This hormone is only of importance in pregnant women.

It has two functions:

1. It contracts the pregnant uterus.

The release of oxytocin from the posterior pituitary gland is one of the factors associated with the onset of the birth of a baby. An intravenous transfusion of synthetic oxytocin is one method of initiating this process [p. 111].

2. It facilitates milk ejection during lactation [p. 111].

THE ANTERIOR PITUITARY GLAND

The anterior pituitary gland (adenohypophysis) consists of glandular tissue. Three types of cell are seen when it is examined histologically after staining with eosin (a red acid stain) and a blue basic stain (see page 34 for the staining of white blood cells in a similar manner). The anterior pituitary cells which take up the eosin stain are termed alpha cells (acidophil cells). The cells which are stained with the basic stain are termed beta cells (basophil cells). The rest of the cells remain unstained and are termed chromophobe cells (G. *khromos* = colour; *phobos* = opposing).

Although it appears that the anterior pituitary gland is controlled by the hypothalamus, there are no nerves running from the hypothalamus to the anterior pituitary gland as there are from the hypothalamus to the posterior pituitary gland. The mechanism by which this control is exercised was solved by the Hungarian medical student Popa. In 1930 he found a portal system running down the pituitary stalk from capillaries in the hypothalamus via portal veins to capillaries in the anterior pituitary (c.f. the portal vein carrying blood from the digestive tract capillaries to the liver).

Experimental evidence is rapidly accumulating to support the hypothesis that each of the anterior pituitary hormones is controlled by a hypothalamic **releasing factor** which passes along this hypothalamo-hypophyseal

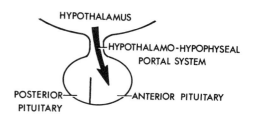

FIG. 119. The anterior pituitary is thought to be controlled by releasing factors which pass from the hypothalamus to the gland via the hypothalamo-hypophyseal portal veins.

portal system and regulates the release of the anterior pituitary hormone into the general circulation [FIG. 119].

Human Growth Hormone (HGH)

Human growth hormone released from the anterior pituitary gland is a hormone which has no specific target organ, but acts on body tissue generally. Its release is under the control of the human growth hormone releasing factor (HGH–RF) from the hypothalamus. This has now been isolated and has been found to be a ten amino acid peptide.

The human growth hormone during childhood stimulates the growth of bone and muscle tissue and by so doing determines stature. Overactivity leads to excessive growth (gigantism). Underactivity leads to retarded growth (dwarfism). It does not appear to be essential for nervous tissue development so that pituitary dwarfs (unlike thyroid dwarfs or cretins) usually have normal mental development. They are just short people. Many circus dwarfs come into this category.

Recent work has shown that injection of human growth hormone will stimulate growth in growth-retarded children, and enable the child to reach a normal stature.

Such treatment is limited at the present moment by the fact that human growth hormone is a 188 amino acid protein which is not yet available in the synthesized form. It is also species specific and this means that growth hormone from an animal cannot be used. The only source is thus from post-mortem specimens of human pituitary glands and as many as 400 glands may be required to treat one child for 18 months.

At the age of 18–24 the epiphyses of long bones fuse with the shafts and then it is not possible to grow any taller. A tumour of the anterior pituitary gland which produces an excess amount of growth hormone later in life will not bring about an increase in stature. The bones of the hands, feet and skull never lose the power of growing and the patient complains that his hands and feet are getting bigger so that his gloves and shoes no longer fit. The overgrowth of facial bone produces facial disfiguration. This condition is termed acromegaly (from the Greek *akron* = extremity).

There is evidence that a low level of human growth hormone is released in the normal adult and that it is playing a part in the regulation of fat metabolism.

PITUITARY TROPHIC HORMONES

The remaining hormones produced by the anterior pituitary are 'trophic' hormones which regulate the activity of other endocrine glands. They are:

(a) Thyrotrophic hormone, alternatively known as the thyroid stimulating hormone (TSH). This hormone controls the thyroid gland.

(b) Adrenocorticotrophic hormone (ACTH) which controls the cortex of the adrenal gland.

(c) Gonadotrophic hormones. These are three in number: follicle stimulating hormone; luteinizing hormone; luteotrophic hormone (prolactin). The hormones control the gonads or sex glands. These are the ovaries in the female and the testes in the male.

Thyroid Stimulating Hormone (TSH), Thyrotrophic Hormone

The thyroid stimulating hormone produced by the anterior pituitary gland acts on the **thyroid gland** in the neck and stimulates the release of the thyroid hormone **thyroxine.** The release of TSH by the anterior pituitary is under the control of the releasing factor TSH-RF from the hypothalamus. TSH-RF is a three amino acid

peptide which has been isolated, synthesized and used for the investigation and treatment of patients.

The blood level of thyroxine acts on the anterior pituitary and hypothalamus to reduce their activity [FIG. 120]. This forms a **feedback system** which tends to maintain a constant level of thyroxine in the blood.

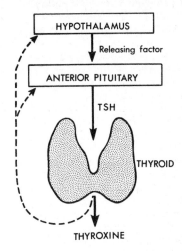

FIG. 120. The thyroid gland is controlled by the thyroid stimulating hormone TSH from the anterior pituitary. The release of TSH is in turn controlled by the TSH-releasing factor from the hypothalamus. Thyroxine inhibits the release of TSH and TSH-RF.

THE THYROID GLAND

The thyroid hormone thyroxine stimulates metabolism. It acts on all the cells of the body and increases the rate at which food is used up and converted into heat and energy.

Thyroxine is an amino acid which contains four iodine atoms. Chemically it is known as tetra-iodothyronine. The corresponding compound with only three iodine atoms is more active and it seems likely that the thyroxine is converted into this three iodine derivative **tri-iodothyronine** before it acts.

The thyroid gland is unique amongst the endocrine glands in that it stores its hormone as a colloid in small vesicles in the gland. The other endocrine glands store their hormones in the cells themselves. This colloid is termed **thyroglobulin.** The thyroglobulin is broken down by a protein-splitting enzyme releasing thyroxine. The hormone thyroxine circulates in the blood attached to plasma protein (α globulin).

The formation of thyroxine by the thyroid gland requires iodine in the diet. Iodine is present in sea-water and areas by the sea where seaweed is used for manure for the crops, have no deficiency of iodine in the diet. In areas remote from the sea a deficiency may exist unless iodine is supplemented in the diet. The most convenient way is to add iodide salts to the table salt.

A swelling of the thyroid gland is termed a goitre. If there is a deficiency of iodine in the diet an iodine deficiency goitre develops. This is associated with under-activity of the gland. A goitre may also be due to a tumour. If this tumour is producing an excessive amount of thyroxine, it will be associated with over-activity of the gland (**thyrotoxic goitre**).

UNDERACTIVITY OF THE THYROID GLAND

If the thyroid gland is underactive (hypothyroidism), the circulating level of thyroxine in the blood will fall, and there will be a reduction in the body's metabolism.

In a child, this reduction in metabolism will slow the developmental processes leading to a stunted, mentally retarded dwarf known as a **cretin.**

In the adult, the condition is termed **myxoedema.** The body temperature is lowered. The heart rate is slower. Brain activity becomes sluggish and in addition there is the deposition of semi-fluid material under the skin that gives the name to this condition (G. *myxa* = mucus; *oidema* = swelling). The skin becomes rough and coarse. The hair thins on the eyebrows and scalp. The face and eyelids become puffy.

Myxoedema can be treated by taking thyroxine by mouth. It is a stable chemical and unlike the hormones which are proteins, it is not destroyed by the digestive juices.

OVERACTIVITY OF THE THYROID GLAND

If the thyroid gland is overactive (hyperthyroidism) the metabolic rate increases. More heat is produced and the skin becomes hot and sweaty. The heart beats at a faster rate, and this increase is maintained during sleep. The sleeping pulse rate may, however, be difficult to determine since the brain is stimulated by the thyroxine and the person becomes a very light sleeper. The in-creased excitability of the cardiac muscle may result in ectopic pacemakers arising in the atria leading to atrial fibrillation [p. 12].

Thyrotoxicosis is the term used when a person with hyperthyroidism becomes clinically ill with the condi-tion. The mental stimulation makes the patients very nervous, irritable and difficult to nurse. Due to the fact that food is converted rapidly to heat, these patients frequently lose weight although their appetite is still good.

Thyrotoxicosis is treated medically and surgically. Drugs such as methimazole and carbimazole prevent the uptake of iodine by the gland, and thus reduce the formation of thyroxine. However, the gland may in-crease in size during this treatment, and if the gland extends into the thorax behind the sternum this glandu-lar enlargement may press on the trachea causing ob-struction.

In an older patient, thyrotoxicosis may be treated by giving radioactive iodine. This is taken by mouth. It enters the blood stream and is selectively taken up by the thyroid gland. The radioactivity destroys the glandular tissue and thus reduces the formation of thyroxine.

The surgical treatment consists of the removal of most of the gland (partial thyroidectomy). The practice is to remove too much, rather than too little, of the gland, since any resultant myxoedema can be readily treated by giving thyroxine by mouth. The gland in thyrotoxicosis is extremely vascular and this makes the arrest of haemorrhage technically very difficult. How-ever, the vascularity can be reduced by giving large doses of iodine during the week prior to the operation. Its mode of action is not known, and it only gives a temporary remission to the thyrotoxicosis. If the opera-tion has to be postponed for any reason the iodine will cease to act and it may be necessary to send the patient home for six months, after which time the iodine treat-ment will be effective once again.

Exophthalmic Goitre

A patient with thyrotoxicosis often has an anxious staring expression. This is due to protrusion of the eye-balls (exophthalmos). It is particularly likely to occur if the thyrotoxicosis is due to overactivity of the anterior pituitary (which might in itself be due to overactivity of the hypothalamus, FIG. 119). The exophthalmos does not appear to be due to an excess of TSH, but to some unidentified substance from the anterior pituitary known as the *exophthalmic substance*.

Thyroxine for Slimming

Thyroid tablets by mouth have been used to treat obesity. If a person has a Calorie requirement of 2,250 Calories per day and eats 2,500 Calories the intake exceeds the requirements, and fat will be deposited under the skin leading to obesity. If the Calorie require-ments are increased to 2,750 Calories per day by taking thyroxine, the intake will now be less than the require-ments and the body fat will be used up. Care must be taken, however, to ensure that the increase in thyroxine in the blood does not lead to atrial fibrillation.

INVESTIGATION OF THYROID ACTIVITY

Since the thyroid gland activity affects metabolism, the determination of the metabolic rate under basal conditions (basal metabolic rate or BMR) gives a guide to thyroid activity [p. 71].

The blood cholesterol level is low in thyrotoxicosis (100 mg. per 100 ml.) and high in myxoedema (600 mg. per 100 ml.). The normal level is 200 mg. cholesterol per 100 ml. blood. Thyroxine lowers the blood choles-terol level by increasing the excretion of cholesterol in the bile.

The plasma thyroxine level is difficult to measure clinically. If facilities are not available the protein bound iodine (PBI) is determined instead and used as an index of thyroid activity. PBI is a measure of thyroxine plus other iodine protein compounds. It does not include the circulating tri-iodothyronine which is not protein bound. The protein bound iodine is high in thyrotoxicosis and low in myxoedema.

The investigation of thyroid activity has been greatly aided by the use of radio-isotopes of iodine. Two are employed, iodine—132 which has a very short life (half-life of 2 hours) and iodine—131 which has a half-life of 8 days. The body cannot distinguish between these isotopes of iodine and normal iodine, and they are all incorporated into thyroxine in proportion to their relative amounts in blood.

Radio-isotopes have the advantage that they can be measured by their radioactivity. An oral dose of radio-active iodine gives a measurable concentration in the neck region four hours later and this can be measured using an external radiation counter. The uptake is usually high in thyrotoxicosis and low in myxoedema.

The briskness of the ankle jerk [p. 131] is related to thyroid activity.

THE PARATHYROID GLANDS

Although the parathyroid glands are not controlled by the anterior pituitary, they will be considered next since they are anatomically and physiologically related to the thyroid gland.

There are four parathyroid glands embedded in the thyroid gland. They produce the parathyroid hormone (parathormone) which maintains the plasma calcium level at 10 mg. calcium per 100 ml. plasma. There is no calcium in the red blood cell, and for this reason the plasma calcium level is usually considered rather than the blood calcium level.

The blood calcium level is 5·5 mg. calcium per 100 ml. blood.

The role of vitamin D in the regulation of plasma calcium level has already been considered [see p. 73].

Overactivity of the Parathyroid Gland

With a parathyroid tumour which is actively producing parathyroid hormone (hyperparathyroidism), the plasma calcium level rises to 20 mg. calcium per 100 ml. plasma. This calcium has come from the bone and is ultimately excreted from the body in the urine. The bones become rarefied and this may lead to spontaneous fractures.

The rarefaction is detected by the reduction in the opacity of the limbs to X-rays when compared with a normal limb. The density of an X-ray film depends on the calcium content of the bone, but it also depends on the exposure time and duration of photographic development. For comparison, it is usual to X-ray the patient's limb and a normal limb side-by-side and to compare the density of the two shadows. Hence a patient with an X-ray which shows two left arms is not a freak, but a case of a suspected parathyroid tumour!

Bone consists of protein which has been strengthened by the deposition of calcium hydroxyapatite. This is a calcium salt composed of calcium hydroxide and calcium phosphate. Under normal conditions bone is being continually broken down and reformed as the mechanical stresses change. New bone is formed by **osteoblast** cells. Old bone is removed by **osteoclast** cells.

Underactivity of the Parathyroid Gland

Underactivity of the parathyroid gland (hypoparathyroidism) leads to a fall in plasma calcium. This fall increases the excitability of the nerves and neuromuscular junctions, leading ultimately to **tetany** if the plasma calcium falls to below 6 mg. calcium per 100 ml. plasma. The spasm of the hands and feet is termed **carpopedal spasm.** In addition there may be **laryngeal spasm.** Increased excitability of the nerve cells in the brain may lead to convulsions.

The increase in nervous excitability may be detected at an early stage by tapping the facial nerve as it crosses the angle of the jaw (**Chvostek's sign of latent tetany**). If the plasma calcium level is low, the face muscles on that side will twitch. Alternatively squeezing the arm or applying a blood pressure cuff will cause spasm of the forearm with extension of the index and middle fingers—*main d'accoucheur* (**Trousseau's sign of latent tetany**). [See hyperventilation tetany, p. 57.]

Hypoparathyroidism may follow a partial thyroidectomy. Three parathyroid glands are frequently removed at this operation and the fourth may be affected by the operation. As a result parathyroid activity is depressed.

The treatment of hypoparathyroidism is to raise the blood calcium. If the calcium is given by mouth, an adequate vitamin D intake will be needed for its absorption and utilization.

Calcitonin

A second hormone which affects plasma calcium was originally thought to be produced by the parathyroid gland. It is now known to be produced mainly, if not entirely, by the thyroid gland. It is termed **calcitonin** or **thyrocalcitonin.** The action of this hormone is to lower the blood calcium by trapping the calcium in bone. Its clinical importance has yet to be evaluated.

THE ADRENAL GLANDS

There are two adrenal (suprarenal) glands situated on the superior aspect of each kidney. Each adrenal gland consists of a central medulla, and an outer cortex.

ADRENAL MEDULLA

The adrenal medulla releases catecholamines (principally adrenaline and noradrenaline) in response to nerve impulses which pass along the preganglionic sympathetic nerves to this part of the gland [FIG. 121]. The catecholamines augment the activity of the sympathetic nervous system [p. 116]. In addition adrenaline mobilizes liver glycogen [p. 67].

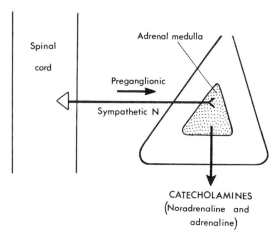

FIG. 121. The sympathetic nerve supply to the adrenal (suprarenal) medulla consists of preganglionic fibres only. There are no postganglionic fibres. Instead the catecholamines, noradrenaline and adrenaline are released into the blood as hormones in response to preganglionic nerve activity.

Phaeochromocytoma

This is a tumour of the adrenal medulla which leads to the over-production of catecholamines. The release of excessive amounts of noradrenaline and adrenaline produce episodes of very high blood pressure [p. 26] Treatment is to remove the tumour surgically.

ADRENAL CORTEX

The adrenal cortex consists of three layers. These are, from outside inwards: zona glomerulosa, zona fasciculata, zona reticularis. The zona glomerulosa produces the hormone aldosterone which facilitates the reabsorption of sodium by the kidney tubules [p. 81]. An excess of this hormone causes sodium (and water) retention, and a loss of potassium.

The control mechanism for the release of aldosterone has not been fully elucidated. It is not under the control of the anterior pituitary. Instead, it appears as if aldosterone is controlled by angiotensin which is formed by the renin released by the kidneys.

The zona fasciculata and zona reticularis of the adrenal cortex makes a series of **corticosteroids** (also known as corticoids or simply steroids). The chief naturally occurring corticosteroid is cortisol (hydrocortisone).

The actions of the hormone cortisol are complex, but they may be summarized as follows:

1. It favours the utilization of protein for the production of heat and energy rather than the use of carbohydrate.
2. It is anti-allergic.
3. It is anti-inflammatory.
4. It has some aldosterone action, that is, it causes retention of sodium (and water) and loss of potassium.

The salt and water retention may lead to oedema and hypertension.

The utilization of carbohydrate for the production of heat and energy is facilitated by **insulin** which enables glucose to enter the cells. Before protein can be utilized in this way, the amino acids are deaminated, and the non-nitrogen part of the molecule used for the production of heat and energy [FIG. 122].

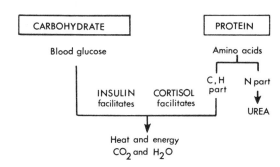

FIG. 122. A high blood cortisol level favours the utilization of protein for the production of heat and energy. The blood glucose utilization is suppressed, leading to a high blood glucose level and glucose in the urine. C = carbon. H = hydrogen. N = nitrogen.

An increase in the **cortisol** level in the blood will reduce carbohydrate metabolism and increase protein metabolism. As a result the blood glucose will remain unused and its level will rise. If the level exceeds the kidney tubular maximum for glucose (blood glucose level in excess of 180 mg. glucose/100 ml. blood) glucose will appear in the urine, leading to **diabetes mellitus.** This form of diabetes mellitus is **insulin resistant** since it is due to an excess of cortisol.

At the same time, body protein will be broken down and used for heat and energy. This applies particularly to recently laid down fibrous and connective tissue. Wound healing will be impaired.

The sodium (and water) retaining properties of cortisol will lead to oedema. The oedema is particularly noticeable in the face which becomes round and full moon-like. A characteristic feature is that the ears are no longer visible when the patient is viewed from the front.

Synthetic corticoids (steroids) have been developed which have less sodium retaining properties than cortisol, but which retain the metabolic, anti-allergic and anti-inflammatory actions. These include prednisolone (derived from cortisol), prednisone (derived from cortisone) and betamethasone. Fluorine derivatives have been found to have a greater anti-inflammatory action than cortisol itself.

Steroid therapy is used in the treatment of a wide range of conditions (including arthritis), allergic diseases (like asthma), and inflammatory skin conditions including inflammation of the eye. It is also of value in the nephrotic syndrome (which is normally associated with a high loss of protein in the urine) and certain blood disorders such as thrombocytopenic (platelet deficiency) purpura. Corticosteroids have been used to suppress the rejection of transplanted organs [p. 87].

Since these corticosteroids lead to the breakdown of body protein, care must be exercised in their use in patients with peptic ulcers (cause exacerbation), tuberculosis (remove fibrous tissue walling off organisms and release tubercle bacilli), and following a recent operation (impair wound healing). The presence of an infection may be masked by the suppression of inflammatory reactions. A **greater amount of corticosteroid is required** during periods of **stress.**

The high blood level of corticosteroids inhibits the release of ACTH from the anterior pituitary and the natural release of cortisol from the adrenal cortex is suppressed. After prolonged therapy the dosage of corticosteroids should be reduced gradually to allow the anterior pituitary to resume its release of ACTH and the adrenal cortex its release of cortisol.

Cushing's Disease

An excess of cortisol due to overactivity of the adrenal cortex gives rise to Cushing's syndrome if it is primarily due to the adrenal cortex itself. If it is the result of an excess of ACTH due to a pituitary tumour (basophil adenoma) the condition is termed **Cushing's disease.**

The sodium and water retention leads to **oedema** (particularly of the face, see above) and to an **increase in blood pressure.** There may be an insulin resistant diabetes with **glycosuria.**

There is an increased deposition of fat on the trunk (but not the limbs) and a characteristic pad of fat at the back of the neck—'a buffalo hump'. The skin bruises easily and shows purple striae with hirsutism in the female. There may also be psychological changes.

The high cortisol level leads to an increase in the excretion of cortisol derivatives in the urine (especially 11-hydroxy compounds).

The treatment is adrenalectomy (or hypophysectomy) with subsequent hormone replacement.

Conn's Disease

An excessive production of aldosterone usually due to a tumour of the zona glomerulosa tissue of the adrenal cortex is termed Conn's disease or **primary aldosteronism.** It is a rare condition associated with muscular weakness and high loss of potassium (and water) in the urine. The subject is therefore thirsty and drinks excessively (polydipsia) and produces a large volume of urine (polyuria).

Addison's Disease

Underactivity of the adrenal gland is termed Addison's disease. It may be due to atrophy of the gland or its destruction by tuberculosis. Both the cortex and medulla are affected.

Sodium and water are lost from the body leading to a weight loss, a low blood pressure, muscular weakness, nausea and vomiting. Catecholamine production is deranged [p. 103] with the formation of the dark pigment **melanin** instead of adrenaline and noradrenaline. The skin becomes pigmented especially in the exposed areas. Since adrenaline plays an important part in mobilizing liver glycogen, when required, to maintain the blood sugar level, episodes of low blood sugar (hypoglycaemia) may occur in this condition. Since an adequate blood glucose level is essential for the normal functioning of the brain, the hypoglycaemia may be associated with mental changes (irritability, confusion and ultimately coma). A patient in an Addisonian crisis may be comatose.

Treatment consists of raising the sodium, water, glucose and cortisol levels (by intravenous infusions if necessary) and then maintaining the patient on daily steroids.

Adrenogenital Syndrome

Tumours of the adrenal cortex may produce *androgenic* hormones (having male sex hormone properties) which lead to virilism in women and precocious puberty in boys. Similar changes may also result from congenital inborn errors of metabolism which derange the production of cortisol by the gland.

MELANOPHORE STIMULATING HORMONE

The pars intermedia which lies between the anterior and posterior parts of the pituitary gland is thought to produce a melanophore stimulating hormone (MSH). Chemically MSH has the same amino acid chain as part of ACTH. It stimulates the melanocytes in the skin to produce the black pigment melanin. The importance of MSH in man has not been fully evaluated but an increase in its production may account for the increased pigmentation seen in Addison's disease, in some cases of thyrotoxicosis, and in pregnancy [p. 110].

18. REPRODUCTION

PITUITARY GONADOTROPHIC HORMONES

The anterior pituitary produces three gonadotrophic hormones which control the activity of the ovaries in the female and the testes in the male. These hormones in the female are the **follicle stimulating hormone (FSH)**, the **luteinizing hormone (LH)** and the **luteotrophic hormone (LTH)** also known as **prolactin**. In the male the same hormones are present, but the luteinizing hormone is referred to as the **interstitial cell stimulating hormone** (ICSH), and the luteotrophic hormone has no known function.

The principal difference is in the release of these gonadotrophic hormones. They are released cyclically in the female (in sequence on approximately a 28-day cycle), but continuously in the male.

OVARIAN HORMONES

The two ovaries, situated in the pelvis, not only produce the female egg cell or ovum. They are also endocrine glands which produce the hormones:

1. Oestrogen.
2. Progesterone.

The female sex hormone oestrogen appears in the circulation at puberty and its presence leads to the development of the secondary sexual characteristics in the female which include development of the breasts and the female distribution of fat and body hair.

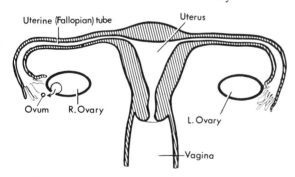

FIG. 123. Diagram showing female reproductive system at ovulation.

Each month, from the menarche to the menopause, one or other of the ovaries produces an ovum, which is discharged at ovulation into the abdominal cavity. It enters the open end of the uterine tube (Fallopian tube) of the same or the other side, and is carried along this tube to the cavity of the uterus [FIG. 123].

If the ovum is not fertilized by a male sperm it only remains in the uterine cavity for 14 days. The endometrium (inner lining of uterus) then breaks down and is shed together with the ovum and a certain amount of blood (usually about 50 ml.) over the course of 3–5 days as the menstrual flow.

The first day of menstruation is a convenient date to note, and it is taken as the first day of the menstrual cycle [FIG. 124]. Menstruation lasts until day 5. Ovulation occurs on approximately day 14, and the next menstruation (and cycle) starts after day 28. Cycles vary in duration. They may be as short as 23 days or as long as 35 days. They may be irregular, but whatever the length of the cycle there is a fairly constant interval of 14 days between ovulation and the next menstruation.

The events in the cycle are controlled by the anterior pituitary gland, and if it is removed, the menstrual cycle ceases. The follicle stimulating hormone (FSH) from the anterior pituitary, stimulates the maturation of one ovum in an ovarian follicle each month. Shortly before ovulation, a surge of luteinizing hormone (LH) appears in the circulation. The combined action of the follicle stimulating hormone and this luteinizing hormone brings about ovulation on day 14.

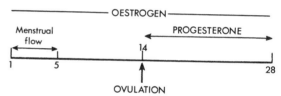

FIG. 124. The menstrual cycle starts with the onset of menstruation (day 1). It lasts, on the average, 28 days. There is an interval of about 14 days between ovulation and the next menstruation.

The cavity left in the ovary by the discharged ovum becomes filled with blood and is converted into a **corpus luteum** (L. = yellow body). It is this corpus luteum that produces the hormone **progesterone.** As its name implies progesterone is a pro-gestation hormone, that is, it prepares the body for pregnancy. Since the corpus luteum is not formed until ovulation has occurred on day 14, progesterone is only present in the circulation in the second half of the menstrual cycle [FIG. 124].

If the ovum is not fertilized, the corpus luteum starts to degenerate about day 26, and progesterone disappears from the circulation. It is the disappearance of progesterone at the end of the cycle that causes the endometrium to break down giving rise to the next menstrual flow.

Ovulation and the formation of the corpus luteum is under the control of the luteinizing hormone (LH) from the anterior pituitary. The release of progesterone from the newly formed corpus luteum is possibly under the control of the third pituitary hormone, the luteotrophic hormone (LTH). The sequential release of FSH and LH may be explained as follows. As the ovarian follicle develops it produces oestrogen and this tends to inhibit the release of FSH and to stimulate the release of LH.

If the ovum is fertilized then the corpus luteum persists, and it becomes a **corpus luteum of pregnancy.** As a result there is no fall in progesterone level and hence no menstrual flow after day 28. The menstrual periods do not start again until after the baby has been born. The progesterone from the corpus luteum of pregnancy (and later in the pregnancy from the placenta) inhibits the cyclical release of gonadotrophic hormones by the anterior pituitary and ovulation ceases.

Oral Contraceptives

The fact that progesterone (and oestrogen) inhibits ovulation is made use of in oral contraception. If an orally active form of progesterone is taken daily from the fifth day of the cycle, the ovulation due on day 14 will not take place, and conception will be impossible. If the progesterone is withdrawn on day 26, the blood progesterone level will fall, and the endometrium will breakdown and bleeding similar to a menstrual period will occur after an interval of about two days. The progesterone is then resumed after an interval of a week, that is, on the fifth day of the next cycle. The interruption in the progesterone level allows this menstrual flow and confirms that the woman is not pregnant. It is not essential for the contraception.

Since the synthetic progesterones have a slight androgenic action (causing growth of hair on the face, etc.), a small amount of synthetic oestrogen is included to cancel out this effect. These two hormones given together appear to suppress principally the release of LH by the anterior pituitary.

In the sequential method of contraception, a high level of oestrogen is maintained by ingesting a synthetic oestrogen from day 5 until after the day on which ovulation should have occurred (i.e. until day 16). Progesterone is then taken for the next 5 days of the cycle and its withdrawal leads to menstrual bleeding two days later. In this method, the high level of oestrogen appears to suppress principally the release of FSH.

TESTICULAR HORMONE

The two testes, situated in the scrotum produce the male sperms. In addition they act as endocrine glands and produce the male sex hormone:

Testosterone

This hormone is made by the interstitial cells of the testes under the control of the interstitial cell stimulating hormone (ICSH) from the anterior pituitary. ICSH is probably identical to LH.

Testosterone appears in the circulation at puberty and brings about the development of the male secondary sexual characteristics, growth of the beard, deepening of the voice, male distribution of hair, and the stimula-

tion of skeletal and muscular growth seen in the male at puberty.

There is no monthly cycle or menopause in the male. The release of ICSH and testosterone continues throughout life although the levels may fall later in life.

The formation of sperms by the testis is probably under the control of FSH from the anterior pituitary.

CONCEPTION

When sexual intercourse takes place the male penis is inserted into the female vagina and as a result of friction between the penis and the vaginal wall a reflex in the male deposits sperms in the vagina.

About 200 million sperms in 3 ml. of semen are released with each ejaculation. These sperms have motile tails and swim through the opening in the cervix into the uterine cavity and along the uterine tubes. Although such a large number of sperms are released, only one sperm fertilizes the ovum.

Ejaculation is a spinal reflex which involves both the autonomic and voluntary nervous systems. The sacral parasympathetic nervous system brings about erection of the penis in response to sexual excitement. If this parasympathetic pathway is damaged (sacral parasympathetic outflow, p. 117), there will be no vasodilatation of the arterioles of the penis, no engorgement of the cavernous tissue and erection can no longer occur.

The stimulation of nerve endings in the penis and surrounding genitalia reflexly stimulates the lumbar sympathetic outflow. This causes contraction of the smooth muscle of the epididymis, vas deferens, seminal vesicles and prostate. As a result seminal fluid is discharged into the urethra. The internal sphincter of the bladder closes so that urine can no longer be voided.

Rhythmical contractions of the muscles surrounding the penis controlled by the voluntary (somatic) nervous system bring about the ejaculation of the seminal fluid.

Following a bilateral lumbar sympathectomy ejaculation is no longer possible although erection of the penis may still occur.

These basic spinal reflexes in the male will be augmented by nerve impulses from higher centres in response to the sexual excitement. Alternatively they may be inhibited by higher centres in states of anxiety, etc.

The generalized increase in sympathetic (and parasympathetic) activity with sexual excitement, which is associated with the release of the hormones noradrenaline and adrenaline from the adrenal medulla, leads to an increase in the heart rate, an increase in the stroke volume, and an increase in blood pressure. [Like exercise the work of the heart is increased and this may put a strain on the diseased cardiovascular system precipitating congestive heart failure.] Respiration is also stimulated and sweating may occur.

It will be remembered that respiration is brought about by the contraction of skeletal (striated) muscle and that the sympathetic nervous system controls smooth and cardiac muscle.

Similar increases in the heart's activity and respiration occur in the female in response to sexual excitement. The blood flow to the external genitalia increases and the secretion of mucus in the vagina and labia facilitates the entry of the penis. But sexual excitement in the female although psychologically desirable, is not essential for conception.

The ovum stands the greatest chance of being fertilized by male sperms as it is passing along the uterine tube. For this reason, sexual intercourse close to the time of ovulation is most likely to lead to conception.

DEVELOPMENT OF THE FERTILIZED OVUM

The **fertilized ovum** starts to develop in the uterine tube and by the time it reaches the uterine cavity it has reached the multi-cell stage. Here it embeds itself in the uterine wall and develops over the course of the next 9 months into a baby. Should its passage along the uterine tube be arrested for any reason, the cell mass will ultimately rupture the uterine tube (**ruptured ectopic**). In exceptional cases the cell mass will develop into a baby in the abdominal cavity.

The baby thus starts as a single cell formed by the entry of a single sperm into the ovum. The tail of the sperm breaks off and the head fuses with the nucleus of the ovum. This single cell now contains all the inherited information which will determine the characteristics of the baby from the colour of its eyes, the colour of its hair, to any possible inborn error of metabolism such as phenylketonuria [p. 67] or galactosaemia [p. 67] or abnormal haemoglobin such as sickle cell anaemia [p. 66].

Genetic Code for Inheritance

The genetical material for inheritance lies mainly, if not entirely, in the nucleus of the cells. This information is transmitted during cell division to the new cells and is thus present in all cells of the body.

The chromosomes in the nucleus consist of DNA which is itself made up of long chains of alternately the pentose sugar desoxyribose and phosphoric acid. The genetical information is carried in the sequence of four different bases which are attached to these sugar molecules. For the synthesis of protein, for example, every three bases in succession forms a codon (or triplet code) which determines which of the twenty amino acids shall be incorporated next in the protein being manufactured.

The fact that only four bases in a triplet code can supply the information for the correct sequence of as many as 20 amino acids can be confirmed with a pack of playing cards. If the playing cards are shuffled, and dealt one by one and the sequence of the suits in every series of three cards noted (hearts–clubs–clubs, diamonds–clubs–spades, etc), it will be found that there are more than 20 alternative combinations (actually 64). It appears likely therefore that there is more than one triplet for some amino acids.

Chromosomes

Human cells contain 46 chromosomes. Two of these are the sex chromosomes which are **X** and **Y** in the male, and **X** and **X** in the female. A baby receives 22 ordinary chromosomes and one sex chromosome from its father and another 22 ordinary chromosomes and a sex chromosome from its mother. Half the baby's characteristics thus come from the mother and half from the father.

It will be noted that half of each parent's characteristics are NOT inherited. It is for this reason that inheritable diseases are not necessarily transmitted to all offspring.

Sex of a Baby

The sex of the baby is determined by the father. Male sperms are of two types, 'the girl-producing sperms' which carry the **X** chromosome and the 'boy-producing sperms' which carry the **Y** chromosome. The ovum contains only the **X** chromosome. If an X-chromosome sperm fertilizes the ovum, the baby will have **XX** chromosomes and will be a girl. If a Y-chromosome sperm fertilizes the ovum, the baby will have **XY** chromosomes and will be a boy.

Although both types of sperms are produced in approximately equal numbers, the sperm carrying the smaller **Y** chromosome is lighter than the sperm carrying the larger **X** chromosome. This may favour the chance of an Y-chromosome sperm reaching the ovum first and account for the fact that 106 boys are born for every 100 girls.

Karyotyping

Chromosomes are not normally visible in the nucleus of a cell. They can only be seen under a light microscope when a cell is dividing and this makes the detection of chromosome abnormalities difficult. It was the discovery that colchicine will stop cell development at the metaphase stage when the chromosomes are visible, that has made their study possible. Even so, it is necessary to grow the body cells in tissue culture before a suitable preparation can be obtained. The development is stopped by colchicine, and hypotonic saline is then used to make the chromosomes swell and disperse. They can then be squashed on to slides. A photograph of the chromosomes is next taken and the chromosomes are cut out from a photographic enlargement. The chromosomes are then paired off according to size into 22 pairs (44 chromosomes) plus the two sex chromosomes. (The main criterion is the relative length of the short arm.) Any additional or abnormal chromosomes can then be detected. This process is termed karyotyping.

Chromosome Abnormalities

Certain pathological states are due to abnormalities in the chromosome pattern of the cells. These may be caused by the failure of either the ovum or the sperm to carry exactly the correct number of chromosomes (22 chromosomes plus one sex chromosome).

Mongolism (Down's syndrome) is usually due to an extra chromosome being transmitted to the baby so that it has a total of 47 chromosomes. An additional chromosome is added to the 21st pair. Some forms of mongolism are due to another abnormality known as translocation. There are the correct number of chromosomes, but chromosome no. 21 has become attached to one of the 13-15 group pairs.

Although **XY** and **XX** are the normal patterns of sex chromosomes in the male and female respectively, other patterns are occasionally found:

    XO (female—Turner's syndrome)
    XXX ('Super' female)
    XXY (male—Klinefelter's syndrome)
    XYY (male)
    XXXXY (male)
    YO (not compatible with life).

In many of these abnormal cases the gonads are rudimentary and the person is sterile. **XXX** and **XYY**, however, are usually fertile.

Chromatin Bodies

Polymorphonuclear white cells in the normal female often show a drum-stick shaped chromatin body attached to the multilobed nucleus. This consists of the **XX** sex chromosomes. Such cells are said to be chromatin positive, and aid in the determination of sex. Chromatin bodies can also be found in the mucosal cells taken from the mouth of a female. More recently fluorescent stains have been used to show up the **XX** chromosomes in hair cut 1 cm. from the root.

Multiple Pregnancies

Should the mother ovulate twice and both ova are fertilized, then twins will develop. These twins are non-identical and may be of a different sex.

Twins can also result when the cells from a single fertilized ovum split into two groups. At this early stage, each group can develop into a complete baby. The twins will be identical. They will be of the same sex and will have the same genetic background since they are both derived from the same ovum and sperm. One result of this is that organ transplants and skin grafts can be made later in life between the twins without rejection. Twins occur in about 1 in 80 of pregnancies. About one third of these are identical.

THE PLACENTA AND PLACENTAL HORMONES

The developing fertilized ovum or embryo develops two membranes, an inner membrane termed the **amnion** and an outer membrane termed the **chorion**. The amnion secretes the **amniotic fluid** in which the embryo develops. This fluid is released with the onset of labour when the membranes rupture.

In the early stages, following implantation, the embryo is completely embedded in the uterine wall and at this stage it receives its food supply via chorionic villi surrounding the embryo which are in the uterine wall. As the **embryo** develops into a **foetus** (pl = foetūs), it bulges into the uterine cavity and the area of contact of the chorion with the uterine wall is reduced to the area of the **placenta.**

The **umbilical cord** connects the foetus, 'floating' (more accurately immersed) in the amniotic fluid, to the placenta which is firmly attached to the uterine wall. Blood is pumped by the foetal heart along the umbilical cord to the placenta. Here it gives off carbon dioxide and waste products to the maternal blood sinuses and returns carrying oxygen and food substances.

Foetal Circulation. Before birth the circulation is different from that after birth. The two sides of the heart in the foetal circulation are in parallel [FIG. 125]. The

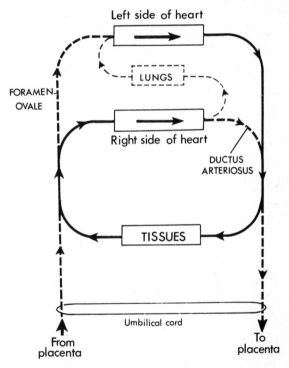

FIG. 125. Circulation before birth.

two sides of the heart share the pumping action and each side pumps about half the blood to the tissues and the placenta. This is possible because of the presence of an opening in the heart between the right and left atria (**foramen ovale**) so that blood returning to the heart via the superior and inferior venae cavae goes to both the right and left sides of the heart.

The lungs are collapsed and unaerated. Very little blood flows through them. Blood which is pumped out of the right ventricle passes via the **ductus arteriosus** to the aorta and joins the output from the left ventricle.

The placenta takes the place of the lungs and oxygenates the blood. The oxygenated blood returns to the foetus by the umbilical vein. It by-passes the liver via the **ductus venosus** and enters the inferior vena cava about 80 per cent. saturated with oxygen. It will, however, be diluted with deoxygenated blood which is returning from the rest of the body.

This arrangement [FIG. 125] is less efficient than that after birth [FIG. 126] and the oxygen tension in the arterial blood of the foetus is comparatively low. This low oxygen tension is partly compensated for by the presence of foetal haemoglobin which is able to become saturated with oxygen at a lower oxygen tension.

Circulatory Changes at Birth. At birth the circulation changes entirely. The two sides of the heart cease to be arranged in parallel [FIG. 125] and become arranged in series [FIG. 126]. The same blood now passes first

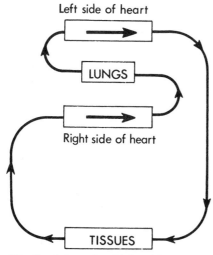

FIG. 126. Circulation after birth. (Compare with FIG. 125.)

through one side of the heart and later through the other side. The umbilical circulation ceases and the blood is now oxygenated by the lungs.

The **foramen ovale** and the **ductus venosus** close shortly after birth. The **ductus arteriosus** closes slightly later, in the first few days of life. It will be noted that in the foetal circulation when the ductus arteriosus is patent (= open) the pulmonary artery blood pressure is as high as the aortic blood pressure $\left(\dfrac{120}{80}\right.$ mm. Hg$\left.\right)$, but that after the ductus arteriosus has closed the pulmonary artery pressure falls to $\dfrac{25}{8}$ mm. Hg. This fall is due to a reduction in the pulmonary vascular resistance. If this reduction does not take place **pulmonary hypertension** persists.

Congenital Heart Disease

Occasionally the ductus arteriosus fails to close completely leading to a **patent ductus arteriosus** which has to be tied off surgically. Failure of closure of the foramen ovale which may be associated with a defect in the atrial septum (**atrial septal defect**) is also repaired surgically.

A congenital defect in the septum between the two ventricles (**ventricular septal defect**) is more difficult to repair since care has to be taken not to damage the bundle of His (atrioventricular bundle) which runs in this septum.

In **Fallot's tetralogy** there is over-riding of the aorta (so that the aorta takes origin from both the left and right ventricles). There is also a ventricular septal defect, pulmonary stenosis and right ventricular hypertrophy. The two sides of the heart are not completely in series and some blood will by-pass the lungs with each circulation. As a result the arterial blood is a mixture of oxygenated and deoxygenated blood. The baby may be visibly cyanosed (blue baby). These defects are corrected surgically.

Placenta as an Endocrine Gland

As well as being an organ of nutrition for the baby, the placenta is an endocrine gland, and releases progesterone, oestrogen and human chorionic gonadotrophic hormone (HCG). Once the placenta has developed, the corpus luteum of pregnancy in the ovary is no longer essential for the maintenance of pregnancy and an ovariectomy (oophorectomy) could be carried out.

The high levels of HCG in the blood lead to its excretion in the urine. The presence of HCG in the urine is used as a test of pregnancy. If the menstrual periods have been regular and intercourse has taken place about the time of ovulation, then the absence of the next period almost certainly indicates pregnancy. The embryo is then only 14 days old. It is when periods have been irregular or absent that pregnancy tests are of value.

The **pregnancy test** commonly employed is the agglutination inhibition test. Gonadotrophin-treated particles (such as latex or even red cells) are added to the urine together with an agglutinating antiserum. Normally agglutination of the particles takes place, but if HCG is present in the urine, it will react with the antiserum and inhibit the agglutination of the particles. Thus pregnancy is indicated by an absence of agglutination.

MATERNAL CHANGES IN PREGNANCY

The uterus increases in size enormously in pregnancy and ultimately reaches the xiphisternum.

The breasts develop and the nipple and areolar area become darkly pigmented. This pigmentation is also seen along the linea alba of the abdomen.

Water is retained later in pregnancy and pregnant women are regularly weighed to check that this water retention is not excessive.

In addition to the high levels of oestrogen, progesterone and human chorionic gonadotrophic hormone found in pregnancy, there is also increased thyroid, parathyroid and adrenal cortex activity.

Morning sickness and a craving for unusual foods are common in early pregnancy. The reason is unknown.

Child-birth

A baby is born by the contractions of the uterus which consists of smooth muscle. Uterine contractions increase in frequency until they are occurring about once every two minutes. After each contraction the uterus relaxes. The cervix of the uterus is dilated by pressure from the baby during this **first stage of labour.** When the cervix has fully dilated the baby is forced from the uterus into the vagina and is born during this **second stage of labour.** The separation and expulsion of the placenta half an hour or so later constitutes the **third stage of labour.** A firm sustained contraction of the uterus after this event minimizes bleeding from the exposed area left by the separation of the placenta. If necessary the uterus may be made to contract further by an intravenous injection of **ergometrine.** The baby suckling at the breast will stimulate the release of **oxytocin** from the posterior pituitary. Although there will be no milk for the baby immediately after birth, it has been shown that allowing the baby to suck at this time decreases the size of the uterus.

The **onset of labour** is not fully understood. The nervous system plays a relatively minor role, since babies have been born by women who were paralysed from the waist down due to spinal injury. The fall in the blood progesterone level due to changes in the placenta and the corpus luteum of pregnancy together with the release of oxytocin from the posterior pituitary are probably the main factors which bring about the uterine contractions.

Uterine contractions can also be started by rupturing the membranes and releasing the amniotic fluid, or by an intravenous infusion of oxytocin or a uterine contracting prostaglandin ($PGE_{2\alpha}$ or $PGF_{2\alpha}$). The oxytocin augments the rhythmical uterine contractions.

LACTATION

Human milk is available to feed the baby a few days after the birth. It is made by the mother's mammary glands from constituents in the maternal blood. The milk contains protein, fat and carbohydrate which will satisfy the baby's requirements for heat and energy as well as for growth and repair of tissue. The baby is able to double its birth weight on human milk alone, although the modern tendency is to introduce additional food supplements at an early age.

Cow's milk has less carbohydrate than human milk. It also has a greater amount of the protein casein which forms a relatively insoluble mass in the baby's stomach after being coagulated by the stomach enzyme pepsin. Human milk protein is mainly lactalbumin. To make cow's milk a suitable substitute for human milk it has to be diluted and sugar added.

The human **mammary gland** consists of about 15 separate milk-producing systems arranged radially around the nipple [FIG. 127]. The milk is produced deep to the surface in the **milk-producing alveoli.** These alveoli lead via a series of branching ducts (like a bunch of grapes) to a main milk duct which opens at the nipple. There are thus 15 openings at the nipple, each leading to its own milk producing alveoli.

The main milk duct has a dilatation just below the surface which is known as the **lactiferous sinus.** Milk which has collected in this sinus is available for the baby who expresses the milk by taking the nipple and surrounding tissue into its mouth and using a sucking champing action. But the baby cannot by this action obtain milk from the deeper milk-producing alveoli.

However, the suckling stimulates the sensory receptors around the nipple and a nervous reflex via the hypothalamus releases oxytocin from the posterior pituitary gland. The oxytocin contracts the myoepithelium surrounding the alveoli and forces the milk forwards towards the nipples. This process is termed **milk ejection.**

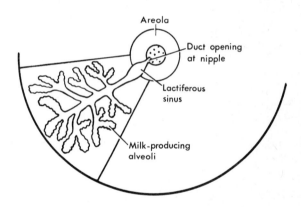

FIG. 127. The human mammary gland consists of 15 milk-producing systems each having a duct which opens at the nipple.

Without an adequate level of circulating oxytocin, the breast may be engorged with milk, but the baby will be unable to obtain an adequate milk supply. Such a condition may be relieved by giving oxytocin.

A similar situation arises in the dairy industry. It is not possible to milk a cow if the oxytocin level in the cow's blood is too low. Injection of oxytocin is employed to bring about 'milk let-down'.

The breast development is rudimentary in the male. In the female, the breasts develop at puberty under the influence of the female sex hormone oestrogen which is released by the ovaries into the blood stream at this time. Much of the enlargement of the breast at puberty is due to the deposition of fat.

As has been seen, progesterone appears in the circulation during the last two weeks of each menstrual cycle and under the combined influence of oestrogen and progesterone the alveoli and ducts start to develop. This causes a slight enlargement of the breasts towards the end of each menstrual cycle. A similar enlargement may occur when synthetic progesterones are taken as oral contraceptives.

The high levels of oestrogen and progesterone which are maintained during pregnancy bring about a marked increase in the alveoli and duct development so that the mammary gland has become an efficient milk-producing organ by the end of pregnancy.

After the birth of the baby the milk alveoli are stimulated to produce milk by the luteotrophic hormone (prolactin) produced by the anterior pituitary.

THE NEW-BORN BABY

A new-born baby is completely dependent on others for its survival. It is provided with a most efficient (and compelling) call system when it requires attention, namely its cry.

A human baby is born before the nervous system has completely developed. As a result the baby is unable to make fully co-ordinated movements or even to change its position to any extent.

The temperature regulating centre is poorly developed and the clothing must be adjusted to prevent over-cooling or over-heating. The newborn has an additional protection against cold in brown body fat, which is metabolized with the production of heat. This fat disappears after a few days of life [p. 91].

19. AUTONOMIC NERVES

ANATOMY OF THE NERVOUS SYSTEM

Spinal Cord Segments

The spinal cord in man develops on a segmental basis. Spinal nerves (forming part of the somatic nervous system) leave on both sides from each segment of the spinal cord. Each spinal nerve has two roots on each side which leave the vertebral canal via the intervertebral foramina between the vertebrae. The ventral (anterior) nerve root conveys the motor fibres whilst the dorsal (posterior) nerve root conveys the sensory fibres [FIG. 128].

The vertebral column consists of 7 cervical vertebrae, 12 thoracic vertebrae, 5 lumbar vertebrae, and 5 sacral vertebrae which are fused to form the sacrum. But there are 8, not 7, cervical nerves (and 12 thoracic nerves, 5 lumbar nerves and 5 sacral nerves). This is because the nerve which leaves between the skull and the first cervical vertebra (atlas) is termed the **1st cervical nerve,** whilst the nerve which leaves between the seventh cervical vertebra and the first thoracic vertebra is termed the **8th cervical nerve.** The **1st thoracic nerve** leaves between the first and second thoracic vertebrae.

Thus in the cervical region the number of the nerves correspond to the vertebra below, whereas in the thoracic, lumbar and sacral regions, the number of the nerves correspond to the vertebra above.

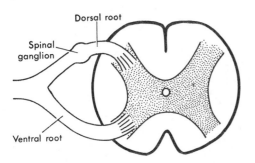

FIG. 128. Bell-Magendie Law. The motor nerves leave via the ventral nerve root. The sensory nerves enter via the dorsal nerve root.

Cerebrospinal Fluid

The spinal cord and brain are covered by three membranes termed the pia mater, arachnoid mater and dura mater. The subarachnoid space, between the pia and arachnoid, is filled with cerebrospinal fluid.

Cerebrospinal fluid (C.S.F.) is formed by the *choroid plexuses* (vascular tufts) in the ventricles (cavities) of the brain. The C.S.F. passes into the subarachnoid space surrounding the brain and spinal cord (via the foramina of Luschka and Magendie) and by forming a sur-

rounding water bath protects the central nervous system from injury. C.S.F. is absorbed by the *arachnoid villi.* These arachnoid granulations are protrusions of subarachnoid space into the venous sinuses.

Lumbar Puncture

The spinal cord is much shorter than the canal in which it lies, and itself only extends down to the 1st lumbar vertebra. The spinal nerves leaving the spinal cord run progressively more obliquely downwards. The lower ones run vertically downwards as the cauda equina (horse's tail), to emerge from their corresponding foramina lower down.

The subarachnoid space extends down to the 2nd sacral segment. A lumbar puncture, to obtain a sample of cerebrospinal fluid for analysis, is carried out below L.1 and above S.2, usually in the space between L.3 and L.4.

TYPES OF MUSCLE

Striated Muscle

In addition to cardiac muscle [CHAPTER 3], there are two further types of muscle in the body. The first type, which is employed in 'voluntary' movement, is termed **striated muscle.** Such muscles are used for skeletal movements such as walking, talking and writing. The respiratory muscles (intercostals and diaphragm) also come into this category.

These muscle cells (muscle fibres) are long and thin, and under the microscope they show cross-striations. Each muscle fibre is supplied by a single motor nerve [FIG. 129 (*left*)]. When nerve impulses pass along these

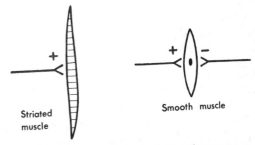

FIG. 129. Striated muscle has a single excitatory nerve supply. Smooth muscle in most parts of the body is doubly innervated with an excitatory and an inhibitory nerve. These come, one from the sympathetic nervous system and one from the parasympathetic nervous system.

motor nerves, the muscle fibres shorten and the muscle contracts. When no nerve impulses are passing along the motor nerves, the muscle relaxes. This relaxed state is termed **flaccidity.**

If the motor nerve to the muscle is interrupted by injury or disease, the muscle supplied becomes paralysed in a relaxed state. This is termed a **flaccid paralysis.**

All motor nerves to striated muscle are part of the **somatic** (or voluntary) nervous system, and are excitatory or plus (+) nerves.

Smooth Muscle

The second type of muscle found in the body is termed **smooth muscle.** The muscle fibres are shorter and fatter and under the microscope they show no cross-striations. These muscle fibres are alternatively referred to as **unstriated** muscle fibres or **plain** muscle fibres. This type of 'involuntary' muscle is found in the digestive tract, air passages, bladder, uterus, blood vessels and controlling the pupil of the eye.

In many parts of the body the smooth muscle is in a state of partial contraction, and there are then two directions in which its length can be changed. The fibres may be made to shorten further, or they may be made to lengthen.

In order to bring about these two changes the smooth muscle has a double nerve supply. One nerve is the **excitatory** or plus (+) nerve which brings about contraction of the muscle, the other nerve is the **inhibitory** or minus (−) nerve which brings about relaxation of the muscle [FIG. 128 (*right*)].

If both nerves are active simultaneously their effects tend to cancel each other out. There will be no change in the degree of contraction of the smooth muscle if the activities in the two nerves are equal. If, on the other hand, the excitatory nerve activity exceeds the inhibitory nerve activity, then the smooth muscle fibre will shorten. Conversely, if the inhibitory nerve activity is greater than the excitatory nerve activity, then the smooth muscle fibre lengthens.

AUTONOMIC NERVOUS SYSTEM

The part of the nervous system which supplies smooth muscle is termed the Autonomic Nervous System (A.N.S.).

The autonomic nervous system is subdivided into:

(*a*) The sympathetic nervous system.

(*b*) The parasympathetic nervous system.

When a smooth muscle fibre has a double nerve supply, one nerve will come from the sympathetic nervous system and the other from the parasympathetic nervous system. As will be seen later, in some parts of the body the sympathetic is the excitatory nerve; in others it is the inhibitory nerve.

In addition to supplying the smooth muscle of the body, the autonomic nervous system also supplies:

(*a*) The heart [p. 28].

(*b*) Secreting glands such as sweat glands [p. 88] and digestive glands [p. 92].

SYMPATHETIC NERVOUS SYSTEM

The sympathetic nervous system is a two-neurone system, that is, the nerve fibre (or neurone) which leaves the spinal cord is not the nerve fibre which arrives at the smooth muscle. The first nerve fibre stops in a sympathetic ganglion. The second nerve fibre starts from the ganglion and runs to the termination [FIG. 130].

The nerve fibre running from the spinal cord to the ganglion is termed the **preganglionic** fibre. It is medullated, that is, each nerve axon is covered by a fatty

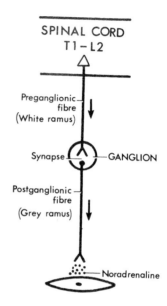

FIG. 130. Sympathetic nervous system. This is a two-neurone system with a synapse between the preganglionic fibre and the postganglionic fibre. The chemical transmitter at the postganglionic termination is noradrenaline. These are adrenergic nerves.

myelin sheath. This white fatty sheath gives the nerve fibres a white appearance. It is referred to as the **white ramus.**

The fibre after the ganglion is termed the postganglionic fibre. This fibre is non-medullated. It has no white fatty myelin sheath, and is grey in colour. It is referred to as the **grey ramus.**

In the ganglion the preganglionic fibre comes into close contact with the postganglionic fibre forming a **synapse.** Nerve impulses in the preganglionic fibre are transmitted across the synapse to the postganglionic fibre.

Sympathetic Outflow

The preganglionic sympathetic outflow is restricted to T.1—L.2. The preganglionic sympathetic fibres have their cells of origin in the lateral horns of grey matter in these segments. The fibres leave the spinal cord, with the motor nerves to voluntary muscle, via the ventral nerve roots [FIG. 131].

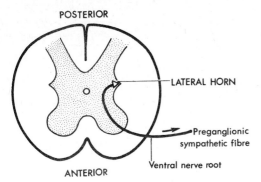

POSTERIOR

LATERAL HORN

Preganglionic sympathetic fibre

Ventral nerve root

ANTERIOR

FIG. 131. The preganglionic sympathetic fibres have their cells of origin in the lateral horns of grey matter of the spinal cord.

Thus sympathetic fibres are only found in the ventral nerve roots of the thoracic and upper lumbar nerves. These preganglionic fibres, however, run to the **sympathetic trunk,** which lies a few centimetres away from the vertebral column on either side [FIG. 132].

The sympathetic trunk extends upwards as far as the superior cervical ganglion which lies in the neck at the level of the angle of the jaw. The sympathetic trunk lies at the back of the thorax and abdominal cavity and extends downwards into the pelvis.

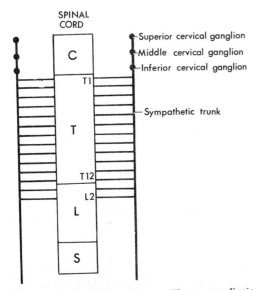

SPINAL CORD

Superior cervical ganglion

Middle cervical ganglion

Inferior cervical ganglion

C

T1

Sympathetic trunk

T

T12

L2

L

S

FIG. 132. Sympathetic outflow. The preganglionic sympathetic outflow is restricted to T.1–L.2 segments of the spinal cord. The postganglionic fibres arise from the sympathetic trunk which runs the whole length of the vertebral column.

Postganglionic fibres arise from ganglia associated with the sympathetic trunk along its entire length. They usually join spinal nerves to run to their destination. Thus the postganglionic sympathetic fibres to the forearm and hand arrive via the three main somatic nerves of the arm (median, radial and ulnar). The postganglionic sympathetic fibres to the head, eye and brain are an exception in that these fibres run to their destination in the outer coat of the carotid arteries and their branches.

The sympathetic fibres to the digestive tract are a further exception. These fibres do not relay in the sympathetic trunk. Instead the preganglionic fibres leave the sympathetic trunk again as the splanchnic nerves which relay in mid-line ganglia in front of the abdominal aorta.

The ganglia, referred to as plexuses, are found around the branches of the abdominal aorta, and are named correspondingly: coeliac plexus (coeliac trunk), superior mesenteric plexus (superior mesenteric artery), inferior mesenteric plexus (inferior mesenteric artery) and hypogastric plexus (hypogastric artery). The postganglionic fibres run from these ganglia to the digestive tract in the abdomen.

Functions of the Sympathetic Nervous System

The sympathetic nervous system is active in states of emotional excitement and stress, that is, states associated with the, so called, 'flight or fight' reaction.

Increased sympathetic activity causes the heart to speed up and the force of contraction of the ventricles to increase. This leads to an increase in blood pressure. In addition the pupils of the eyes dilate, the air-passages increase in diameter, sweating occurs and the contraction of the arrectores pilorum muscles causes the 'hair to stand on end' and goose-pimples to form. In addition movements of the digestive tract will be reduced.

In such a state there will probably also be a stimulation of breathing brought about by an increase in respiratory muscle activity as a result of the action of higher centres on the respiratory centre [p. 55], but it must be remembered that the respiratory muscles are 'striated muscle' and are therefore not under the control of the autonomic nervous system. (The muscle in the bronchi and bronchioles is, however, smooth muscle which is relaxed by the sympathetic nervous system, hence the dilatation of the air-passages referred to above.)

Sympathetic Chemical Transmitter

When the nerve impulses reach the end of the postganglionic nerve fibre, they bring about the release of the chemical transmitter **noradrenaline** [FIG. 130]. This chemical transmitter bridges the very small gap between the end of the nerve and the muscle fibre. This noradrenaline acts on the smooth muscle and brings about either a contraction or a relaxation, depending on whether it is an excitatory or inhibitory nerve ending [FIG. 129 (*right*), p. 113].

The noradrenaline is rapidly removed after release (mainly by a re-uptake into the nerve) so that the muscle is able to respond to future nerve impulses.

The **adrenal medulla** endocrine glands release noradrenaline, and its methyl-derivative adrenaline, as hormones when nerve impulses pass along the preganglionic sympathetic fibres leading to the gland [FIG. 121, p. 104]. Chemically, noradrenaline and adrenaline are *amines* of the benzene derivative *catechol*, and if one does not wish to differentiate between them, they may be referred to collectively as **catecholamines.**

When catecholamines are released from the adrenal medulla they will arrive via the blood stream at the sympathetic nerve terminations and augment the effect of the local sympathetic nerve activity. They thus have a sympatheticomimetic action (i.e. they mimic the sympathetic).

Sympathetic Overactivity

Local overactivity of the sympathetic nervous system leads to excessive vasoconstriction which will impair the blood flow to that part of the body. If widespread there will be generalized vasoconstriction which will lead to hypertension. Excessive activity to the sweat glands may lead to profuse sweating of, say, the palms of the hands.

Tumours of the adrenal medulla, which are actively producing catecholamines, mimic bouts of excessive sympathetic overactivity which are associated with a very marked rise in blood pressure. Such tumours are termed *phaeochromocytomas*.

A person with overactivity of the sympathetic nervous system can be treated by surgically interrupting the sympathetic pathways. Such an operation is termed a **sympathectomy.**

Ganglionic blocking agents are drugs which block the transmission at the synapse in the ganglion, and thus prevent nerve activity in the preganglionic fibre being transmitted to the postganglionic fibre. Hexamethonium is an example of such a blocking drug. These drugs provide an alternative to sympathectomy. Unfortunately, however, the parasympathetic nervous system also contains ganglia. These ganglionic blocking agents may also block the parasympathetic nervous system which will have undesirable side-effects such as blurring of vision, and a reduction in the motility of the digestive tract.

Alpha and Beta Blocking Agents

The same chemical substance *noradrenaline* is released from nerves that bring about contraction of smooth muscle, and nerves which bring about relaxation of smooth muscle. The difference in effect is due to a difference in the type of receptor found in the muscle cell [FIG. 133].

If the receptor responds to noradrenaline by bringing about a contraction of the muscle fibres, it is termed an alpha receptor. If, on the other hand, the action of the noradrenaline on the receptor is to make the muscle relax, then it is termed a beta receptor.

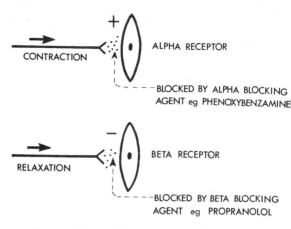

FIG. 133. Sympathetic receptors. Alpha receptors are usually associated with contraction of smooth muscle, whilst beta receptors are associated with relaxation. The sympathetic receptors in the heart (cardiac muscle), however, behave like beta receptors.

Drugs which prevent the action of noradrenaline on the alpha receptor are termed **alpha blocking agents.** An example of such a drug is *phenoxybenzamine* (dibenyline).

Drugs which prevent the action of noradrenaline on beta receptors are termed **beta blocking agents.** Such drugs include *propranolol* and *oxprenolol*.

The **sympathetic fibres** to the **heart** which bring about an increase in heart rate and an increase in the force of contraction act on **beta receptors.** This is rather surprising but we must remember that we are dealing here with cardiac muscle and not with smooth muscle. The increase in heart rate brought about by an increase in sympathetic activity such as in emotional excitement can be prevented by giving the drugs *propranolol* and *practolol* which will block these receptors. To distinguish, cardiac receptors are sometimes termed beta 1, whilst smooth muscle relaxation receptors are termed beta 2.

Monoamine Oxidase Inhibitors

Sympatheticomimetic drugs, related chemically to adrenaline and noradrenaline, are used in cough medicines. Substances with a sympatheticomimetic action are found in certain foods. They are all monoamines. In the body monoamines are destroyed by the enzyme **monoamine oxidase.** If the monoamine oxidase is removed, their action will be enhanced.

Monoamine oxidase inhibitors, which are used for the treatment of depression, potentiate sympatheticomimetic amines such as *ephedrine, amphetamine* and *tyramine*. Tyramine is found in cheese. If the destroying agent is inhibited, cheese will cause hypertension, palpitations and may lead to heart failure. For this reason patients on monoamine oxidase inhibitors are given a list of foods to be avoided. The list includes broad beans, cheese, meat extracts, yeast extracts, yoghurt and red wine. A hypertensive episode in such a patient, resulting from the ingestion of food containing a sympatheticomimetic substance, can be treated by giving blocking agents.

Supersensitivity

If postganglionic sympathetic nerves are depleted of noradrenaline by a drug such as *bethanidine*, the receptors become super-sensitive to circulating noradrenaline. A similar supersensitivity to the transmitter (noradrenaline or acetylcholine) occurs after denervation in smooth muscle, skeletal muscle and some glands.

PARASYMPATHETIC NERVOUS SYSTEM

The parasympathetic nervous system has a craniosacral outflow, that is, the fibres originate in the cranial nerves and from the sacral region of the spinal cord. The cranial nerves which contain parasympathetic fibres are III, VII, IX, X. The tenth cranial nerve or vagus is the principal parasympathetic nerve in the cranial outflow. The sacral outflow is from S.2, 3, 4.

The parasympathetic, like the sympathetic, is a two-neurone system, but in most cases the postganglionic

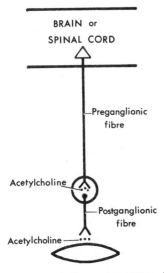

FIG. 134. Parasympathetic nervous system. This is a two-neurone system with a craniosacral preganglionic outflow. The chemical transmitter at the postganglionic termination is acetylcholine. These are cholinergic nerves. Acetylcholine is also the transmitter at the ganglion in both the parasympathetic (above) and the sympathetic [Fig.130] nervous systems.

fibre is very short, and the ganglion and this fibre often lie in the organ supplied.

Thus the vagus nerve which supplies the heart with parasympathetic fibres is the preganglionic part of the two-neurone system; the ganglion and the postganglionic fibres lie in the cardiac muscle.

When a nerve impulse reaches its termination [FIG. 134] it brings about the release of the chemical transmitter **acetylcholine**.

Unlike the adrenal medulla and the sympathetic nervous system, there is no endocrine gland in the body which augments the activity of the parasympathetic nervous system by releasing acetylcholine as a hormone.

The acetylcholine released may bring about contraction, or it may bring about relaxation of the muscle. As yet, the parasympathetic receptors have not been subdivided like the sympathetic, although the acetylcholine may bring about either of the two actions.

Both types of acetylcholine activity are blocked by the drugs **atropine** and **hyoscine** (scopolamine).

Like noradrenaline, acetylcholine is removed very rapidly after its release, this time by the enzyme cholinesterase.

The drug *carbachol* mimics the action of the parasympathetic nervous system (parasympatheticomimetic or cholinomimetic drug). It is often used in place of acetylcholine to augment the activity of the parasympathetic nervous system.

Functions of the Parasympathetic Nervous System

In most parts of the body the action of the parasympathetic nervous system is the opposite of that of the sympathetic nervous system. It slows the heart, lowers the blood pressure, constricts the pupils, and constricts the air passages. In addition the parasympathetic nervous system speeds up the digestion of food and plays an important part in defaecation and micturition. It is the 'emptying mechanism' of the body.

The parasympathetic nervous system stimulates the production of saliva [p. 92], gastric juice [p. 93] and pancreatic juice [p. 94] and increases the motility of the digestive tract, at the same time relaxing the sphincters.

The parasympathetic nervous system is a more discrete system than the sympathetic, that is, its activity, at any given time, may be restricted to certain organs or parts of the body.

COMBINED ACTION OF THE SYMPATHETIC AND PARASYMPATHETIC NERVOUS SYSTEMS

Since the smooth muscle in the body usually has a double nerve supply, there are two ways in which the activity can be modified. Thus a muscle can be made to contract further by augmenting the activity of the excitatory nerve or by inhibiting the action of the inhibitory nerve. By analogy, if two tug of war teams are equally matched then one side will win if that side pulls harder, or if the other side does not pull quite so hard.

Size of Pupils

The size of the pupil of the eye depends on the balance between the activity of the sympathetic nervous system which is tending to dilate the pupil, and the activity of the parasympathetic nervous system which is tending to constrict the pupil so as to give a pin-point pupil [FIG. 135]. Normally both divisions of the autonomic nervous system are active, and the pupil will have an intermediate size depending on the intensity of the light.

If it is desired clinically to dilate the pupil in order to carry out an internal inspection of the eye, then either the sympathetic nervous system activity must be augmented, or the parasympathetic activity must be reduced. It is the latter way that is usually adopted, and homatropine drops (a short-acting derivative of atropine) are applied to the eye, in order to eliminate the parasympathetic activity. The sympathetic is now unopposed and the pupils dilate.

Atropine is obtained from the plant *Atropa belladonna* (deadly nightshade). The name belladonna (= beautiful lady) dates from the time when women found it fashionable to dilate their pupils by putting the extract from

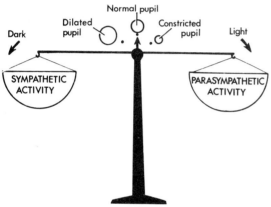

FIG. 135. Size of pupil. Balance between sympathetic and parasympathetic activity determines the size.

this plant in their eyes. Without knowing it, they were paralysing the parasympathetic pupil-constrictor nerves, leaving the sympathetic pupil-dilator nerves unopposed.

Dilatation of the pupils occurs in the dark and in emotional excitement due to the increase in sympathetic activity.

Bronchospasm

The air passages contain smooth muscle. This smooth muscle has a sympathetic and parasympathetic nerve supply. The sympathetic nervous system brings about dilatation of the bronchi (bronchodilatation), whereas the parasympathetic nervous system brings about constriction of the bronchi (bronchoconstriction). With both systems active the bronchi have an intermediate size.

Should the parasympathetic be overactive, or should the bronchi be constricted by the release of histamine, following an antigen-antibody reaction, then bronchoconstriction will occur. This bronchospasm is present in asthma.

In order to relieve this condition it is necessary to either augment the sympathetic activity, or to decrease the parasympathetic activity. An injection of adrenaline or noradrenaline will bring about a generalized increase in sympathetic activity, but in order to produce a local response only, these substances may be given by inhalation. **Isoprenaline** is a derivative of noradrenaline that is particularly active on these beta receptors (causing relaxation). Isoprenaline sprays are frequently employed in the treatment of bronchospasm. Care should be taken because of the harmful effects of excessive dosage with such sprays; due to their sympatheticomimetic action an excessive stimulation of the sympathetic nervous system will affect the heart and may lead to ventricular fibrillation [p. 12].

In addition the parasympathetic may be depressed by atropine, or similar drugs, but these will have a general effect in depressing the whole parasympathetic nervous system.

If the bronchospasm is an allergic manifestation (such as hay fever) then antihistamine drugs and cortisol may be employed to suppress the antigen-antibody reaction.

Digestive Tract

The vagus (parasympathetic) supplies the digestive tract and digestive glands from the oesophagus, stomach, duodenum, jejunum and ileum through to the caecum, ascending colon and part of the transverse colon. The rest of the transverse colon, descending colon, sigmoid colon, rectum and anal canal are supplied by the sacral parasympathetic. The sacral parasympathetic also supplies the bladder and external genitalia. The sympathetic supply is via the splanchnic nerves [p. 115].

The parasympathetic nervous system increases the motility of the digestive tract, and stimulates the production of digestive juices. The sympathetic, on the other hand, reduces the motility of the digestive tract. Thus too much parasympathetic activity will lead to a very fast passage of the food through the intestines, which may lead to malabsorption of the food. Too little parasympathetic activity or too great sympathetic activity will, conversely, lead to slowing of the movement along the digestive tract and possibly to *paralytic ileus*. Such a condition is treated by augmenting the parasympathetic activity by giving a derivative of acetylcholine such as *carbachol*.

Premedication before an Anaesthetic

When a patient is unconscious or is anaesthetized, he can no longer swallow. As a result any saliva produced may be inhaled into the lungs. To prevent such an occurrence during anaesthesia the parasympathetic supply to the salivary glands, which bring about salivation, is blocked by giving a premedication containing atropine or hyoscine. Atropine is employed in children and old people, but since it tends to act as a mental stimulant, it is usually replaced in adults by hyoscine (scopolamine). Hyoscine tends to depress the brain centrally and leads to forgetfulness.

These drugs, which block the parasympathetic nervous system, also reduce the motility of the digestive tract and reduce the likelihood of vomiting. Hyoscine in small doses is used as an anti-seasickness, anti-car-sickness remedy.

The premedication will also include a sedative-tranquillizing drug.

Defaecation

The sacral parasympathetic nervous system plays an important part in defaecation. It causes contraction of the rectum, and relaxation of the internal anal sphincter. After the first year or so of life, the external sphincter, which is under the control of the brain via the somatic nervous system, must also be relaxed before defaecation takes place. Should the brain be damaged or the spinal cord injured, then the basic parasympathetic reflex will be re-established, and whenever the rectum becomes filled with faeces, defaecation will occur.

Micturition

The bladder is composed of smooth muscle under the influence of the sympathetic and the parasympathetic nervous systems. The parasympathetic causes the bladder muscle to contract, the internal sphincter to relax, and micturition to occur. There is an external sphincter under voluntary control which develops its activity in the first years of life. So that, like defaecation, micturition only occurs when the external sphincter is relaxed. Should this voluntary control be lost, then the parasympathetic will bring about micturition whenever the bladder is full.

The sympathetic nervous system relaxes the bladder muscle, and allows the bladder to fill. It also contracts the internal sphincter.

As urine passes into the bladder, the pressure in the bladder increases, and then, after a time, the bladder wall relaxes so the pressure drops slightly. This stepwise increase in the bladder pressure continues until the pressure is sufficient to reach consciousness when the desire for micturition results.

Micturition may be aided in patients who have insufficient parasympathetic activity by injections of carbachol.

Sweat Glands

The sweat glands are innervated by sympathetic nerves only. There are no parasympathetic fibres to sweat glands. However, the sympathetic fibres release acetylcholine instead of noradrenaline and sweating is blocked by atropine.

Blood Vessels

The blood vessels of the body have smooth muscle in their coat which is supplied by the sympathetic nervous system. This sympathetic nerve activity is acting for the most part on alpha receptors so that the blood vessels contract [see vasoconstriction, p. 18]. The blood

vessels in the salivary glands and external genitalia only, have parasympathetic vasodilator nerves in addition to sympathetic vasoconstrictor nerves.

SUMMARY OF AUTONOMIC NERVOUS SYSTEM ACTIVITY [See TABLE 6]

The sympathetic nervous system is highly active in states of emotional excitement. It is often referred to as the 'fight or flight' system. Its activity under these conditions is augmented by the release of catecholamines from the adrenal medulla. The parasympathetic has the opposite action, and the increase in parasympathetic activity during sleep brings about such changes as the slowing of the heart.

The parasympathetic nervous system is the emptying mechanism of the body. It is employed in the movement of food along the digestive tract. It is employed in defaecation, and in micturition.

TABLE 6

Organ Supplied	Sympathetic Activity	Parasympathetic Activity
Pupil of eye	Dilates	Constricts
Air passages, bronchi and bronchioles	Dilates	Constricts
Salivary glands	—	Salivary secretion and dilatation of blood vessels
Heart	Speeds up Increases force of ventricular contraction	Slows
Digestive tract	Reduces motility	Increases motility
Sphincters of digestive tract	Constricts	Relaxes
Rectum	Allows filling	Empties Relaxes internal anal sphincter
Bladder	Allows filling	Empties Relaxes internal sphincter
Blood vessels	Vasoconstriction	Nil (except salivary gland and external genitalia—vasodilatation
Sweat glands	Sweat	Nil

Parasympathetic activity is restricted to the trunk and skull. There are no parasympathetic fibres in the arms or legs.

Paralysis of the cervical sympathetic on one side leads to Horner's syndrome (a unilateral constricted pupil, drooping of the upper eyelid (ptosis), enophthalmos and an absence of sweating).

20. MOTOR AND SENSORY NERVES

MOTOR NERVES

A cross-section of the spinal cord shows a central H-shaped area which is grey in colour surrounded by an outer area which is white in colour. The grey matter has two anterior horns and two posterior horns [FIG. 136]. The grey matter is made up of the nerve cells

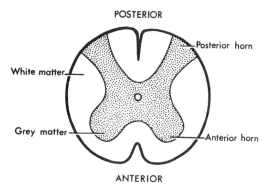

FIG. 136. Cross-section of the spinal cord. A section of the spinal cord shows a central area of grey matter (nerve cells) surrounded by white matter (nerve fibres).

whilst the surrounding white matter consists of nerve fibres.

The cells of origin of the motor nerves are found in the anterior (ventral) horns of grey matter. The nerve fibres which arise from these anterior horn cells leave the spinal cord via the ventral (anterior) nerve roots. These motor nerves, which are also termed **efferent nerves,** supply the **striated (voluntary) muscles.**

MOTOR UNITS

A striated muscle fibre only contracts when nerve impulses are passing along its motor nerve [FIG. 137]. Every muscle fibre needs a nerve supply in order to contract, but if a count is made of the number of muscle fibres and the number of anterior horn cells, it is found that there are many more muscle fibres than anterior horn cells. Every anterior horn cell must therefore innervate many muscle fibres. In the leg muscles as many as 200 muscle fibres may share a single anterior horn cell. In the extrinsic muscles of the eye, where very little power is needed to rotate the globe, as few as 5 muscle fibres are supplied by one anterior horn cell.

Each anterior horn cell gives rise to a single axon which leaves via the ventral nerve root [FIG. 138]. When this axon reaches the muscle supplied, it branches and supplies the group of striated muscle fibres [FIG. 139]. This motor neurone and all the muscle fibres supplied by the same anterior horn cell constitute a **motor unit.**

The motor unit forms the basis for voluntary movement. Each anatomical muscle contains many thousands of motor units, but movements are carried out in terms of motor units rather than in terms of anatomical muscles.

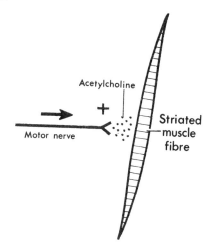

FIG. 137. Striated muscle fibre. The excitatory nerve to striated muscle releases acetylcholine at the nerve termination.

When the anterior horn cell sends a nerve impulse along the motor nerve, every muscle fibre in the motor unit contracts. The individual fibres do not respond on their own as long as the motor nerve is intact.

If the motor nerve is cut, the motor unit is paralysed, and plays no further part in voluntary movement. Under this condition only, individual fibres may twitch giving rise to *fasciculation.*

Voluntary movement is graded by the rate at which the anterior horn cells discharge. If an anterior horn

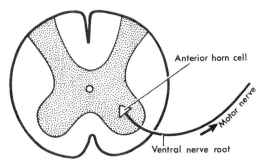

FIG. 138. Anterior horn cell. Motor nerves have their cells of origin in the anterior horn cells in the grey matter of the spinal cord. The nerves leave via the ventral (anterior) nerve roots.

cell is silent (discharge frequency zero) its motor unit is relaxed. If all the motor units in an anatomical muscle are relaxed, the muscle is **flaccid**.

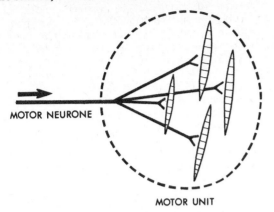

FIG. 139. Motor units. Each motor neurone supplies a number of striated muscle fibres which contract and relax together. They constitute a motor unit.

If an anterior horn cell discharges very slowly, say one impulse per second, the motor unit will twitch with each nerve impulse [FIG. 140]. The upper tracing shows the electrical changes in the nerve (electroneurogram). A spike or action potential corresponds to each nerve impulse. The lower tracing shows that shortly after the arrival of the nerve impulse, the muscle makes a short sharp contraction. The contraction is termed a **muscle twitch**.

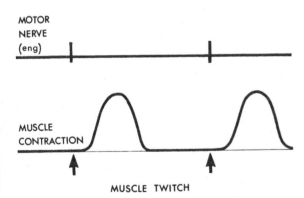

FIG. 140. A muscle responds to each nerve stimulus (upper trace) with a single contraction. This short contraction is termed a twitch. (eng = electroneurogram.)

If the frequency of discharge of the anterior horn cell increases to 20 impulses per second, more continuous contraction of the motor unit is obtained, but the contraction shows a superimposed flutter or tremor [FIG. 141]. This state of tremulous contraction is termed a **clonus**.

When the discharge frequency of the anterior horn cell has increased to more than 50 impulses per second,

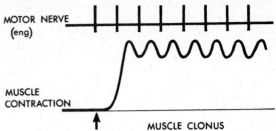

FIG. 141. A muscle responds to a series of nerve stimuli at a low repetition rate (10–30 per second) (upper trace) with a tremulous contraction known as a clonus (lower trace). (eng = electroneurogram.)

an even more powerful steady contraction is obtained [FIG. 142]. This state of sustained contraction is termed a **tetanus**.[7]

This behaviour of a motor unit has been deduced from the behaviour of an anatomical muscle when its motor nerve is electrically stimulated. The preparation usually employed to demonstrate these responses is the sciatic nerve—gastrocnemius muscle preparation of the

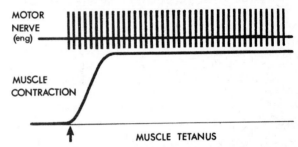

FIG. 142. A muscle responds to a series of nerve stimuli at a high repetition rate (50–200 per second) (upper trace) with a sustained contraction known as a tetanus (lower trace). (eng = electroneurogram.)

frog [FIG. 143], but twitches, clonus and tetanus may also be demonstrated in man by stimulating the motor point of flexor digitorum superficialis in the forearm and recording the flexion of the ring finger.

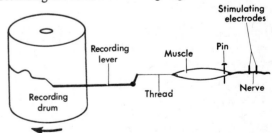

FIG. 143. Nerve-muscle preparation. The frog sciatic-gastrocnemius is used. The nerve is stimulated electrically. The muscular contraction is recorded on a revolving drum (kymograph).

[7] Tetanus is also the name given to an infection (lock-jaw) caused by the tetanus bacillus. The word *tetanus* should not be confused with *tetany* [pp. 57 and 103] which is carpopedal spasm as a result of overventilation or hypoparathyroidism.

In the frog nerve-muscle preparation, the sciatic nerve is stimulated electrically and the resulting contraction of the gastrocnemius muscle recorded either by the movement of a lever with a writing point across a revolving drum (isotonic recording) or electronically using a strain gauge (isometric recording). If a muscle is allowed to shorten the contraction is termed **isotonic**. If it is allowed to develop tension, but not shorten, by pulling against a very strong spring, the contraction is termed **isometric**. Most voluntary actions are a mixture of the two. Pulling an object towards you is an example of an isotonic contraction. Pulling at an immovable object is an example of an isometric contraction.

An electrical stimulus of sufficient strength is employed to bring in all the motor units. As the frequency of stimulation increases, the change from single twitches, to clonus and finally to tetanus can be demonstrated.

Grading of Muscular Contractions

Using a muscle such as biceps it is possible to make a feeble contraction or a powerful contraction. The power is determined by the number of motor units in this muscle which are employed. As more and more motor units are brought into play, the power of the voluntary movement is increased. The force of contraction of each motor unit itself will depend upon the rate at which nerve impulses are being sent to it, that is, the discharge rate of the anterior horn cell.

The force of contraction of an individual motor unit can be increased from complete relaxation if the anterior horn cell is silent, to a series of twitches, if the anterior horn cell is discharging at a low rate of 1 to 5 impulses a second, to a clonus if it discharges at a rate of 10, 20, 30 impulses a second, and finally to a tetanus at a discharge of 50–200 impulses per second.

The fastest rate of discharge of anterior horn cells employed in voluntary movement is about 200 impulses/ second. This rate of discharge is only used when making very powerful contractions.

Asynchronous Discharge of Anterior Horn Cells

An anatomical muscle does not twitch when the discharge rate to the motor unit is low (of the order of one impulse per second) because the anterior horn cells fire asynchronously. The contractions of the neighbouring motor units are thus out of step, and whilst one is contracting, the surrounding units are relaxing. With thousands of motor units in an anatomical muscle twitching asynchronously a steady contraction results.

It is traditional for soldiers marching across a bridge to break step, on the grounds that if they all marched in step, oscillations might be set up in the bridge causing it to collapse. If they are out-of-step, there is only a steady pressure on the bridge.

Should the anterior horn cell firing become synchronized in man, a tremor will result. A patient with a **tremor** is thus a patient in whom the anterior horn cells supplying the muscle group are firing synchronously and not asynchronously. (L. *syn* = together, *chronus* = time, *a* = absence of.)

All the motor units in voluntary muscles are seldom, if ever, used at their maximum capability. The nerve block appears to be in the motor cortex of the brain which does not involve all the possible anterior horn cells to their maximum capabilities. The superhuman strength in an emergency gives some idea of the full motor power that is available. There is thus a great reserve of muscle power which is not normally used.

A patient who has had poliomyelitis, may have lost a large number of anterior horn cells as a result of this virus infection. It is important that these patients should be encouraged to make the best use of the motor units that remain.

One reason for having a crowd at a sports' contest is to encourage the competitors to bring in the maximum number of motor units and thus produce the best performance. After a prolonged athletic performance the competitor may suffer from fatigue. Even though the competitor has collapsed on the ground the nerves and muscles still respond when stimulated electrically. The fatigue lies principally in the brain.

THE NERVE CELL

The generation of a nerve impulse is associated with changes in the nerve cell. FIGURE 144 shows a nerve cell with its cell body, axon and dendrites. The long

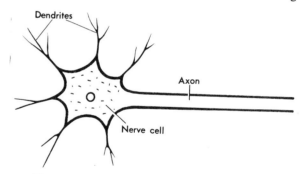

FIG. 144. A neurone. This consists of a nerve cell and its processes (axon and dendrites). The Nissl granules in the cell are absent around the origin of the axon.

process running from a nerve cell to the motor unit is termed the **axon**. The nerve cell plus its processes is termed a **neurone**.

This nerve cell and its processes behave like a small electrical battery. The resting voltage inside the nerve cell is −70 mV.

1 millivolt (mV) is one thousandth of a volt.

This voltage may be represented on a graph [FIG. 145]. To show that the voltage is negative, it is placed below the baseline. The term 'potential' may be used as an alternative to 'voltage'. Thus we may say that the **resting potential** of a nerve cell is −70 mV.

The inside of the nerve cell, like that of any other cell of the body, contains potassium salts. These salts give rise to potassium ions (K^+). The nerve cell is surrounded by tissue fluid. The principal salt in this tissue fluid is

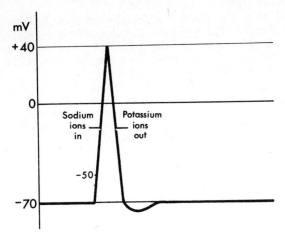

FIG. 145. The resting voltage (potential) in a nerve cell is minus 70 millivolts. An excitatory neurone synapsing with this cell tends to make this voltage less. When the voltage reaches a critical level of minus 50 millivolts, the permeability of the membrane alters, sodium ions rush in and a nerve impulse is generated. The cell voltage changes rapidly to plus 40 millivolts. Potassium ions then come out and the cell voltage returns to minus 70 millivolts again. The whole process takes less than one thousandth of a second (=1 millisecond).

sodium chloride which gives rise to sodium ions (Na⁺).

There are, thus, potassium ions inside the nerve cell and sodium ions outside the nerve cell.

In the resting state sodium ions are prevented from entering the nerve cell by the 'sodium pump'. This is the name given to the metabolic processes whereby any sodium which enters a cell is immediately excreted again [FIG. 146].

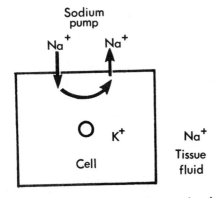

FIG. 146. The cells are kept free of sodium ions by a metabolic process, which extrudes sodium ions, termed a sodium pump.

An analogy is the revolving door of a hotel. If any undesirable character tries to enter via this door, he can easily be expelled again by the hall porter giving a slight push so that as the door revolves he is forced out into the street again.

A nerve impulse is a transient event and for a very short period of time the cell membrane is altered so that it allows sodium ions to pass in. When sodium ions enter the cell, they take with them positive charges which change the voltage inside the cell from −70 mV to +40 mV [FIG. 145]. This process is termed depolarization. Immediately afterwards an equal number of potassium ions leave the cell bringing the voltage back to its initial value of −70 mV.

The alteration in voltage from −70 mV to +40 mV is a change of 110 mV or approximately one tenth of a volt. This sudden change in voltage is termed an action potential.

On a slow time scale a series of action potentials resemble a series of spikes [FIG. 147 (left)]. On a faster time scale the way in which the voltage changes can be seen [FIG. 147 (right)]. A recording of the activity passing along a nerve is termed an electroneurogram. Each spike or action potential corresponds to a nerve impulse.

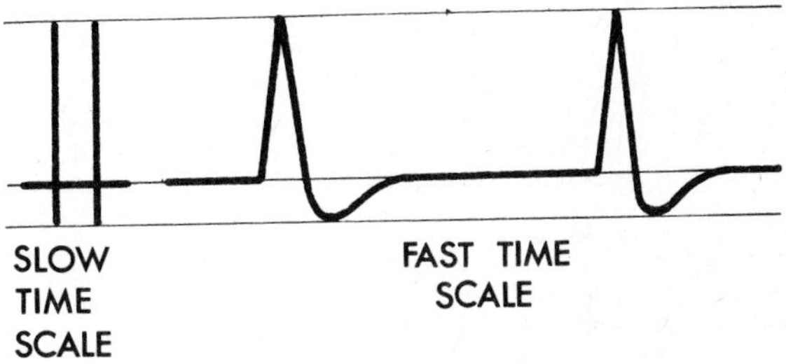

SLOW TIME SCALE   FAST TIME SCALE

FIG. 147. Action (spike) potentials. The voltage changes associated with the passage of nerve impulses along nerves are called action potentials (or spike potentials). Two nerve impulses are shown on a compressed time scale (left) and an extended time scale (right). These recordings are termed electroneurograms.

The alteration in the state of the membrane is propagated along the nerve. If an alteration takes place in the membrane in the nerve cell, there is a progressive change with 'sodium in' and 'potassium out' all the way along the nerve.

Each nerve impulse in both motor and sensory nerves is associated with sodium entering and potassium leaving. With each nerve impulse only a very small amount of sodium passes in, and only a very small amount of potassium comes out. This amount is small compared with the total amount of sodium and potassium present. During the resting phase there is a gradual expulsion of the sodium which has entered the nerve cell. If this were not so, over the course of a lifetime, so much sodium would enter the nerve cells that the nerves would no longer conduct nerve impulses.

A nerve impulse lasts about 1 millisecond, that is, one thousandth of a second, during which the whole process of the sodium moving in, and the potassium coming out takes place. This gives a theoretical limit of 1,000 impulses per second as the maximum number of impulses that a nerve could conduct in one second. In practice, as has been seen, about 200 impulses a second is the maximum rate of discharge of the anterior horn cells.

Conduction Velocity

The nerve impulses are propagated at a very fast rate in the large diameter nerve fibres. The largest nerve fibres in the body, which have a diameter of 20 microns, have a conduction velocity of 120 metres per second. These are myelinated (medullated) nerves, that is, they have a fatty myelin sheath. The gaps in the sheath which occur every millimetre along the nerve are termed **nodes of Ranvier**. The nerve impulse passing along such a nerve does not pass continuously all the way down the axon, but jumps from one node to the next (or to a node further down the nerve) [FIG. 148].

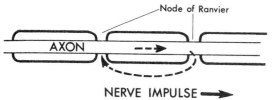

NERVE IMPULSE ➝

FIG. 148. Saltatory conduction. The nerve impulse travelling along a medullated (myelinated) nerve leaps from node to node instead of passing continuously along the axon. The reason for this is that a local electric current (dotted arrows) circulates from the point that the nerve impulse has reached (left-hand node of Ranvier) through the next node returning in the extracellular fluid. This current changes the membrane permeability and hence generates a nerve impulse at the right-hand node of Ranvier.

The nerve impulse thus 'leap-frogs' down the nerve. The latin word for 'to leap-frog' is saltare, and this type of nerve conduction is called **saltatory conduction**.

Some nerve fibres such as those conveying pain are only one micron in diameter, and are non-medullated. They have a conduction velocity of only a few metres per second.

The human body is about 2 metres long. Nerve fibres which conduct nerve impulses at a rate of 120 metres/second, send nerve impulses from the feet to the brain in about one-sixtieth of a second. Nerve impulses passing along non-medullated nerves will take $\frac{1}{2}$ to 1 second. The presence of these two types of nerves with different conduction velocities is confirmed whenever the big toe is placed in the bath and the water is too hot. The toe is withdrawn rapidly using the medullated nerves and there is a delay before the maximum degree of pain is felt.

NEUROMUSCULAR JUNCTION

The junction between the motor nerve and the muscle fibre supplied is termed the **neuromuscular junction.** There is a small gap between the end of the nerve and the motor end-plate of the muscle fibre [FIG. 149]. This gap is bridged by the release of very small amounts of acetylcholine every time a nerve impulse arrives. This acetylcholine passes across the gap and is taken up by receptors on the motor end-plate of the muscle fibre. The acetylcholine alters the membrane permeability of the motor end-plate so that sodium rushes in producing

a change in voltage which is termed an **end-plate potential.** The depolarization spreads from the motor end-plate to the whole of the muscle fibre (propagated action potential), calcium ions are released and the muscle then contracts.

The acetylcholine diffuses away and is destroyed shortly after its release by the enzyme **cholinesterase**. This enables the next nerve impulse to be effective. **Anticholinesterases** oppose the action of cholinesterase. **Myasthenia gravis** which is associated with great muscular weakness is a condition associated with a disorder of the neuromuscular junction.

Muscle Relaxants

The acetylcholine released at the neuromuscular junction is the same substance as that released at parasympathetic terminations [p. 117], but the muscle receptors are different in the two cases. The acetylcholine receptor of the **parasympathetic** nervous system is **blocked by atropine and hyoscine.** The acetylcholine receptor of the **skeletal muscle** is not blocked by atropine but is **blocked by muscle relaxant drugs.** The first muscle relaxant drug introduced into medicine was the

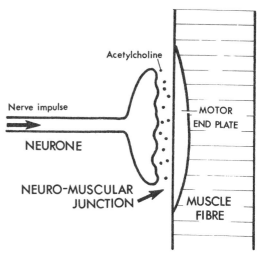

FIG. 149. Neuromuscular junction. The small gap between the nerve and muscle is bridged by the release of acetylcholine. The acetylcholine is destroyed shortly after its release.

American-Indian arrow poison **curare.** They used curare to paralyse the animals which they caught for food. The derivative of curare which is commonly employed as a muscle relaxant is **tubocurarine.** Curare does not prevent the release of acetylcholine but it stops it acting on the motor end-plate. *Flaxedil* acts in a similar manner.

Drugs such as **succinylcholine** also block neuromuscular transmission but they do so by keeping the motor end-plate in a state of polarization. Succinylcholine is destroyed only very slowly by the enzyme **pseudocholinesterase.**

Before the introduction of muscle relaxant drugs, the **depth of anaesthesia** in an abdominal operation had to be increased until the abdominal muscles relaxed. In some cases the level of anaesthesia needed was close to the lethal dose of the anaesthetic. With muscle relaxant drugs very light anaesthesia may be used. The muscle relaxation is achieved by blocking the neuromuscular junctions. One disadvantage to the use of muscle relaxant drugs is that the patient may stop breathing due to a paralysis of the respiratory muscles. It has been seen [p. 46] that respiration is brought about by striated (voluntary) muscles. Artificial respiration may have to be employed for the duration of the operation, and continued until the effects of the muscle relaxant drug have worn off.

Electromyogram

The electrical activity produced by a group of motor units may be recorded by means of surface electrodes placed over the skin adjacent to the muscle mass. To record activity from single motor units, a concentric needle electrode (the size of a fine hypodermic needle) is inserted into the muscle. Since the voltage recorded is only one ten-thousandth of a volt, transistor or valve amplification is needed before the signal is large enough to be recorded on a pen recorder or displayed on a cathode ray oscilloscope. The activity recorded is the electrical activity in the muscles as it becomes depolarized. The record is termed an **electromyogram** (EMG). The electrical activity in a nerve is termed an **electroneurogram** (ENG). For every action potential in the motor nerve there is an action potential in the motor unit supplied.

kilograms recorded on a spring balance, more and more motor units are brought into play (with progressively higher frequency of discharge to each unit and this greater activity shows as 'more ink on the paper').

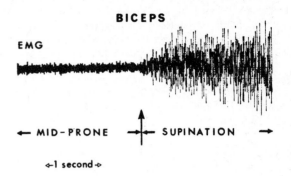

FIG. 151. Electromyogram of biceps muscle as a supinator of the forearm recorded using surface electrodes. At the arrow, the forearm is supinated from the mid-prone position against resistance. Note the increase in biceps activity.

FIGURE 151 shows how an electromyogram recording can be used to confirm that the biceps is not only a flexor of the elbow joint, but also a supinator of the forearm. The tracing shows the activity in biceps when the forearm is held in the mid-prone position. At the arrow the forearm is supinated (thumb outwards) against resistance. Note the increase in the activity of biceps. Biceps has been termed the 'wine-waiter's' muscle since it is used to put the corkscrew in and then pull the cork out (provided that one is right-handed!).

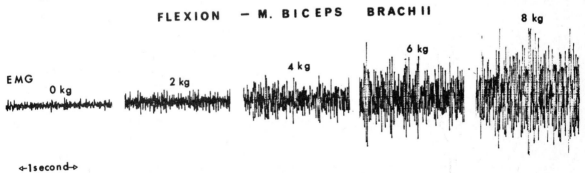

FIG. 150. Electromyogram of biceps muscle recorded using surface electrodes. The muscle is being employed as a flexor of the elbow joint pulling against a spring balance. As the tension recorded on the spring balance in kilograms is progressively increased, more and more units of biceps are brought into play and the amplitude of the electromyogram increases.

FIGURE 150 shows the electromyogram of biceps brachii muscle recorded using surface electrodes applied to the upper arm. Such a technique records activity in a number of motor units close to the recording electrodes. As voluntary contraction of this muscle is increased as indicated by tensions of 0, 2, 4, 6 and 8

SENSORY NERVES

Sensory or afferent nerves bring information into the central nervous system from the receptors in the skin and deeper structures.

There are four different types of skin sensory nerves.

These nerves have receptors in the skin which are sensitive to:

1. Touch.
2. Pain.
3. Heat.
4. Cold.

From deeper structures, in addition to nerves conveying pain, there are nerves which give information about position of joints in space and the degree of contraction of muscles. These are termed proprioceptor nerves. The proprioceptor nerves giving information about the position of joints in space originate from receptors in the joint capsules. Information concerning the degree of contraction of muscles is obtained from the muscle spindles.

Sensory nerves have their cell of origin in the spinal ganglia which lie outside the spinal cord in the dorsal nerve roots [FIG. 152]. As the name implies, the cell of origin is the cell from which the nerve develops. During embryonic development all the nerves start as nerve cells. The nerve fibres grow out from these cells of origin. This cell retains its importance throughout life and should the cell of origin die, then the entire nerve fibre dies.

The spinal ganglion should not be confused with the sympathetic ganglion [p. 113]. Unlike the sympathetic ganglion, there is no synapse or relay point at the spinal ganglion.

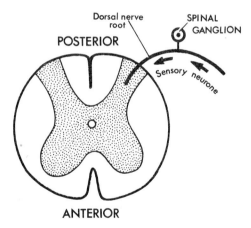

FIG. 152. Sensory nerve roots. Sensory nerves have their cell of origin in the spinal (dorsal root) ganglion. They enter via the dorsal (posterior) nerve roots.

Sensory Nerve Pathways

Nerve fibres carrying the sensation of touch enter the spinal cord via the dorsal (posterior) nerve roots [FIG. 153, pathway (1)]. They run up in the posterior column of white matter of the spinal cord to the **medulla**. In the medulla this nerve synapses with a second nerve which crosses to the other side of the body and runs up to the **thalamus.** In the thalamus a third neurone originates

which conveys the sensory information to the **sensory cortex** which lies on the outer surface of the brain.

There are thus three neurones involved in conveying sensory information from the touch receptor in the skin on one side of the body to the sensory cortex on the other side.

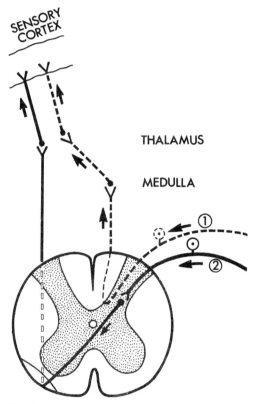

FIG. 153. Sensory nerves travel by two pathways to the thalamus and sensory cortex.

Pathway (1) is taken by fine touch and proprioception.

Pathway (2) is taken by pain and temperature (and some touch fibres).

Sensory Cortex. The outer surface of the cortex of the brain is convoluted and shows a large number of elevations and depressions [FIG. 154]. Each elevation is termed a gyrus (plural = *gyri*), and each depression is called a sulcus (plural = *sulci*). In most brains there is a well-marked sulcus which separates the frontal lobe from the parietal lobe [FIG. 155]. This is termed the **central sulcus** or *fissure of Rolando*. Immediately posterior to this central sulcus lies the **postcentral gyrus.** This is the site of the sensory cortex.

The whole body is represented in the sensory cortex in 'an upside down fashion' [FIG. 155]. The head area is found at the lower end of the gyrus, the feet at the top, and the arms and trunk in between.

Thus if one touches a table with the index finger of the right hand, the information arrives at an area of

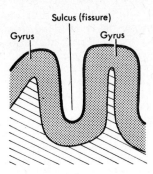

FIG. 154. Convolutions of outer surface of brain. The outer surface of the brain shows sulci and gyri. The grey matter of the cortex lies in this outer layer.

the sensory cortex which is situated about three inches (7·5 cm.) above the left ear.

The same route as that taken by **touch** is taken by **propioception.**

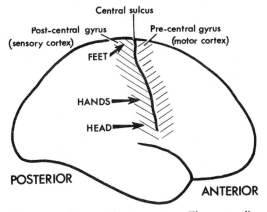

FIG. 155. Sensory and motor cortex. These areas lie on either side of the central sulcus (fissure of Rolando). The body is represented in an upside-down fashion in these areas.

Pain and **temperature** fibres take a different pathway to reach the thalamus and sensory cortex of the other side of the body. The pain and temperature fibres enter by the dorsal nerve root, but the nerve fibres then stop. The cells of origin of these fibres are in the spinal ganglia as in the case of the touch and proprioceptor fibres. The second neurone with which the first neurone synapses in the spinal cord crosses to the other side and runs up in the anterolateral columns of white matter on the other side to the thalamus [FIG. 153, pathway (2)]. These nerve fibres have already crossed and there is, therefore, no decussation in the medulla. A third neurone conveys the information from the thalamus to the sensory cortex.

There are thus two different pathways to the thalamus and sensory cortex. These are shown diagrammatically in FIGURE 156.

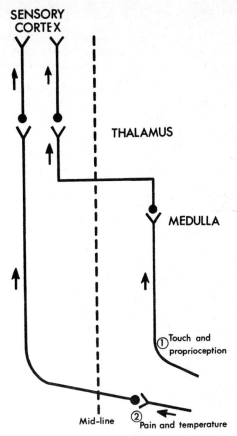

FIG. 156. The two principal sensory pathways shown in diagrammatic form.

Nerve Tract Nomenclature. Nerve tracts are named according to the place where they start and the place to where they are going. Thus the fibre tracts running from the spinal cord to the thalamus are termed the spinothalamic tracts. The fibre tracts running from the thalamus to the sensory cortex are termed the thalamocortical tracts.

The spinothalamic tracts carrying pain fibres are severed to relieve intractable pain.

Body Image and Localization of Sensation

The nerves conveying touch, pain, temperature, etc. transmit this information in terms of nerve impulses. These nerve impulses are identical for all nerves of the same size irrespective of the type of sensation carried. It is not possible by studying the action potentials in a nerve to deduce what type of information is being conveyed. There are, of course, differences in the size of nerve fibres and differences in the velocities of conduction, but it is probably the place in the brain to which the nerve fibres run, that determines the type of sensation experienced.

In the early years of life we develop a **body image** and relate the sensory information that is arriving at the brain to this image. Thus, if someone treads on your toe you know which foot it is, and which toe it is, by relating the sensation to the body image.

The conclusions reached by relating sensory information to the body image may occasionally be erroneous. This is most likely to occur when the circumstances are unusual.

The body image may be confused by crossing the index and middle fingers of one hand. A pencil rubbed between the two fingers often feels like two pencils. Alternatively, two pencils touching opposite sides of these two fingers may feel like one.

If a nerve is stimulated by pressure at some point along its route, the sensation felt is usually assumed to be coming from the receptor at the distal end.

If one had a private telephone line one would assume that all calls were coming from the far end. If someone tapped in on the telephone line and made a call it would be natural to assume that this call was coming from the telephone at the far end.

A protruded intervertebral disc may press on a nerve root and send nerve impulses into the spinal cord. The patient does not complain of a disc pressing on the nerve at the root level. Instead he complains of pain in the area supplied by the nerve, for example, in the case of the lumbar disc down the back of the leg. This is an example of a **referred pain.**

The diaphragm has a nerve supply, the phrenic nerve, which comes from C.3, 4, and 5. The **diaphragm** is poorly represented on the body image. An abscess under the diaphragm may cause pain sensation to enter the spinal cord via nerve roots C.3, 4, 5. This pain is often referred by the patient to the skin area supplied by C.3, 4, 5, that is, the skin over the shoulder or the shoulder joint.

The referred pain from **ischaemia** of the **heart** (angina pectoris) is a tight gripping pain round the chest which extends down the inside of the arm. The pain has been referred to upper thoracic segments. The first thoracic nerve, which is part of the brachial plexus, supplies the skin on the inside of the arm; the other upper thoracic nerves supply the chest wall via the intercostal nerves.

With regard to the **digestive tract,** there is usually accurate localization down the **oesophagus.** Once the food enters the stomach there is no further localization. Pain from the intestinal tract is usually referred to the region of the umbilicus.

One of the most distressing types of referred pain is that following an amputation. Stimulation of the cut ends of the nerves may give rise to the sensation of a phantom limb. Pain is felt in a limb that is no longer present. It may take a long time for the body image to be modified.

Thus the information given by a patient about the localization of a pain may be unreliable if the patient is experiencing a referred pain.

PROPRIOCEPTION

The accuracy of the proprioceptive information concerning the position of joints in space may be tested by extending the arm at shoulder level, closing the eyes and then bringing the hand round to touch the nose with the index finger. This is the **finger-nose test.** A more difficult test is to place the index finger of one hand at a given position in space (say behind the back or behind the neck) and then to take the index finger of the other hand to the same position.

Proprioception in the lower limbs may be tested by running the heel of one leg along the anterior surface of the other leg with, once again, the eyes shut.

The ability to detect passive movement may be tested by moving the subject's big toe upwards or downwards and asking the subject to state the direction of movement. The receptors lie mainly in the joint capsule. The toe should be held by the sides to prevent alterations of pressure on the dorsal and plantar sides of the toe giving away the direction of movement. Since there are only two possible answers (up or down) a subject who has lost all proprioception would, on the average, get 5 correct answers out of 10 by purely guessing, so 5 out of 10 right means no proprioception! A person with normal proprioception gets 10 out of 10 correct.

Proprioception is used in every-day life. It is used, for example, when eating. In this case, the 'proprioception' is extended beyond the hand to the end of the fork or spoon so that its position in space is 'computed' by the brain and the mouth is opened just as it arrives to put the food in. Walking involves the use of the proprioceptors in the legs. When walking very little clearance is allowed between the foot and the ground but this distance is carefully judged. How small this clearance is, is only realized when one trips over an unnoticed slightly elevated pavement stone.

A normal person can keep his balance with the eyes shut, using the proprioceptors in the legs. A person who has lost proprioception in the legs does not know where the ground is relative to his legs and he has a stamping gait. He can maintain his balance using his eyes, but he seldom goes out at night time because he falls over in the dark. Such a proprioceptor loss occurs in **tabes dorsalis** which is a late manifestation of syphilis. A patient with this condition gives a typical history of having fallen over when the soap got in his eyes when he was washing his face. The soap made him shut his eyes.

Maintenance of Upright Posture

When standing a person is maintaining a *dynamic* rather than a *static* equilibrium. He is not balancing on his feet like a rigid table (static equilibrium) but is swaying very slightly the whole time. As he tends to fall forwards the muscles at the back of the legs are stretched. The muscle spindles are stimulated and the stretch reflex [p. 130] brings about a contraction of these muscles. He then tends to fall backwards. This stretches the muscles at the front of the legs. The muscle spindles in these fibres are now stretched, and the stretch reflex pulls him forwards again. The balance is thus maintained by the alternate contraction and relaxation of the flexors and the extensors. At the same time there will be a similar swaying movement to the two sides. If consciousness is lost, he slumps to the ground.

VESTIBULAR APPARATUS

The vestibular apparatus consists of two parts, the **otolith organs** and the **semicircular canals.** The otolith organs are the **saccule** and **utricle.** There are three semicircular canals at right-angles to each other. The vestibular apparatus of the inner ear is poorly developed in man and it plays only a minor role in the maintenance of the upright posture. **Disorders of the vestibular apparatus** are often associated with **vertigo** or giddiness. Together the vestibular apparatus gives information about the position of the head in space (otolith organs), and information about any change in speed or acceleration (semicircular canals). It gives no information about the speed itself. Thus one may be travelling at 700 m.p.h. in a jet plane and have no idea of the speed at which one is travelling. In a motor car it is the acceleration and deceleration that is readily noticed.

The otolith organs are very important for orientation when swimming under water, since under these conditions the feet are not touching the ground and there will be no orientating proprioceptor information arriving at the brain from the legs. Nor will vision be of much assistance in determining the direction of the surface (see p. 135 for vision under water). Anyone who has a vestibular disorder should not go swimming.

Otolith Organs

The saccule and utricle consist of modified hair cells. Embedded in the hairs are chalk particles known as otoliths [FIG. 157]. These chalk particles are heavy

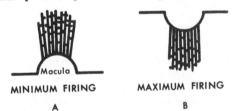

MINIMUM FIRING   **MAXIMUM FIRING**

A   B

FIG. 157. Otolith organ giving information about the position of the head in space. Sensory nerves in the macula are stimulated maximally when hair cells are pointing downwards [B] and minimally when pointing upwards [A]. This depends upon the position of the head.

enough to be pulled on by gravity. In the base of the hairs are nerve endings which respond maximally when the chalky particles are pulling on the hairs. The maximum stimulation occurs when the hairs are pointing downwards [FIG. 157 (B)].

In the upright posture, the utricles are pointing upwards whilst the saccules are pointing outwards and sideways [FIG. 158]. It may be noted that **U** stands for **utricle** and **upwards**, **S** stands for **saccule** and **sideways.** The saccule is stimulated maximally by putting the

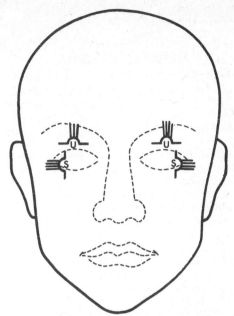

FIG. 158. Diagram to show direction of otolith organs with reference to the skull. S=saccule. U=utricle.

head on one side. The utricles are stimulated maximally when standing on one's head.

Semicircular Canals

By having three semicircular canals at right angles, an acceleration in any direction can be detected.

Each semicircular canal is filled with fluid [FIG. 159].

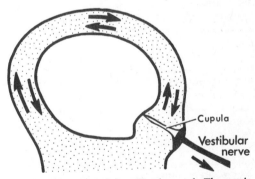

FIG. 159. Structure of a semicircular canal. The cupula of the crista acts as a movable partition which swings when the fluid moves with respect to the canal. Since the fluid has inertia, the crista will be stimulated by any change in rotational speed, i.e. acceleration.

This fluid has inertia and is left behind when the head rotates suddenly. It also carries on moving when the head suddenly stops rotating. In both cases the fluid moves relative to the canal itself. The semicircular canal is blocked at the crista by a movable partition known as the cupula which is made of gelatinous material embedded in hair cells. The fluid movements move the partition and stimulate nerve endings at the base of the hair cells. This sensory information is interpreted as acceleration.

RETICULAR FORMATION

Running through the brain stem from the medulla right up to the cortex is a network termed the reticular formation. It consists of a large number of neurones and a large number of synapses. All sensory information coming into the central nervous system passes to the reticular formation. The reticular formation arouses the cerebral cortex. Certain of the sensory information is sent to consciousness.

Drugs which block the reticular formation will prevent sensory information reaching consciousness. They are termed anaesthetics. The activity of the reticular formation is depressed during sleep.

CORTICAL CONTROL OF MOTOR NERVES

Motor Cortex

The motor cortex lies on the surface of the brain, anterior to the central sulcus, in the **precentral gyrus** [FIG. 155, p. 127]. Like the sensory cortex, the body is represented in an upside down fashion with the head at the inferior end of this gyrus and the feet at the superior end.

Electrical stimulation of the exposed motor cortex brings about the movement of voluntary muscles on the other side of the body.

The cortical areas have been numbered. The motor cortex is area 4. The sensory cortex, discussed above, occupies areas 1, 2, and 3.

The grey matter in the brain is on the surface, unlike the spinal cord where the grey matter is in the central H-shaped area.

The **corticospinal tract** (older name pyramidal tract) originates in the motor cortex. It runs down by the side of the thalamus through a region termed the **internal capsule**. In the medulla the corticospinal tract crosses to the other side, and then continues down in the lateral columns of white matter of the spinal cord to supply anterior horn cells.

When corticospinal fibres reach the spinal segment that they are to supply, the fibres run to the anterior horn cells. In an animal such as the cat, there is a small internuncial neurone between the end of the cortico-spinal fibre and the anterior horn cell but it is absent in primates and man. Each anterior horn cell gives rise to the nerve fibre which runs to the motor unit [FIG. 160].

There are thus two motor neurones involved in the pathway for voluntary movement. The first neurone runs from the motor cortex of the other side to the spinal segment. The second neurone is the anterior horn cell and the motor nerve.

The pathway through the internal capsule may be interrupted by a haemorrhage from the lenticulostriate arteries. These are fragile arteries which may rupture if the blood pressure becomes too high. Such a **cerebral**

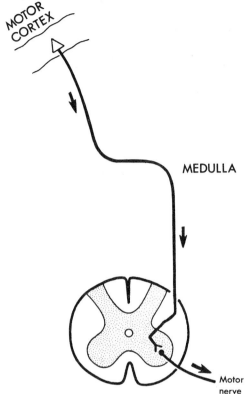

FIG. 160. The pathway for voluntary movement. Two nerves are involved. The pathway from the motor cortex to the spinal cord (corticospinal tract or pyramidal tract) crosses to the left side in the medulla.

vascular accident will interrupt the pathway for voluntary movement, and paralyse muscles on the other side of the body.

A thrombosis of cerebral blood vessels will also lead to paralysis since the blood vessels supply the nerves and the nerves will not function if their blood supply is interrupted. [The paralysis of the ulnar nerve which occurs after sleeping on one's arm is due to an interruption of the blood supply to this nerve by pressure.]

STRETCH REFLEX

Muscle spindles are sensory receptors (in the shape of a spindle) which lie alongside the striated muscle fibres in a voluntary muscle. They are stimulated by stretching.

When a muscle is stretched, nerve impulses from a muscle spindle pass into the spinal cord and act on the anterior horn cells that are supplying the adjacent motor units [FIG. 161]. The anterior horn cells send nerve impulses along the motor nerves and the motor units contract. This constitutes the **stretch reflex**. The pathway [FIG. 161] is termed a **monosynaptic reflex arc** since only one synapse is involved.

Stretch reflexes are found in all the muscles of the body. If any muscle is stretched, it will promptly con-

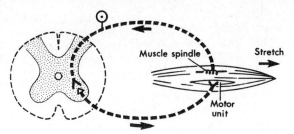

FIG. 161. Stretch reflexes. When a muscle is stretched nerve impulses pass from the muscle spindle into the spinal cord to act on the anterior horn cells which supply neighbouring motor units. These nerve impulses cause the muscle to contract. This forms the basis of jerk reflexes.

tract. It is technically difficult to stretch some muscles but in certain parts of the body the muscles can be stretched easily. For example, the quadriceps muscle of the thigh has a tendon which passes round the lower end of the femur and is inserted into the tibia [FIG. 162]. In the middle of this tendon is a sesamoid bone, the patella. By tapping the tendon between the patella and

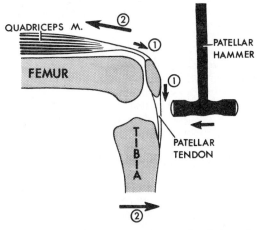

FIG. 162. Knee-jerk. The patellar tendon is tapped with a rubber-covered patellar hammer. This exerts a sudden pull on quadriceps muscle (1). A monosynaptic reflex, originating from the muscle spindle, causes the muscle to contract (2). This contraction is transmitted through the patellar tendon to the tibia. The knee extends and the leg kicks forward.

the tibia with a rubber-covered patellar hammer a pull is exerted on the quadriceps muscle [FIG. 162, (1)]. The monosynaptic reflex causes the muscle to contract [FIG. 162, (2)]. The resultant contraction extends the knee and kicks the foot forward. This reflex is termed the **knee-jerk.**

It should be noted that the sensory receptors stimulated by tapping the patellar tendon are the **muscle spindles** lying in the **quadriceps muscle** remote from the point of impact. Receptors in the patellar tendon itself play no part in this reflex, since an identical response is obtained after these Golgi tendon organs have been infiltrated with a local anaesthetic.

Another muscle that can be easily stretched by tapping its tendon is the gastrocnemius-soleus muscle. This is inserted into the calcaneum at the ankle via the Achilles tendon [FIG. 163]. The response to tapping

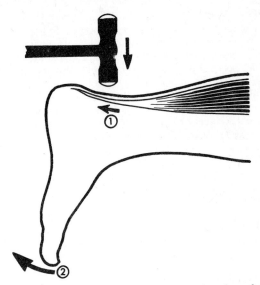

FIG. 163. The ankle-jerk. The combined tendon of gastrocnemius and soleus (tendo achillis) is tapped by a patellar hammer. The muscle is stretched (1), and reflexly contracts causing the foot to plantar-flex (2).

this tendon [FIG. 163, (1)] is a contraction of the gastrocnemius muscle and plantar flexion of the foot [FIG. 163, (2)]. This reflex is called the **ankle-jerk.** The ankle-jerk is influenced by thyroid activity. In thyrotoxicosis [p. 102] this reflex is brisk. In myxoedema, on the other hand, the reflex is sluggish and there is a delay before the muscle relaxes.

The reflex may also be initiated by rapidly dorsi-flexing the foot to stretch the muscle. If the response is a clonic contraction the condition is termed **ankle clonus.**

Other stretch reflexes include the biceps-jerk and the triceps-jerk in the arm, and the jaw-jerk.

It has already been seen that the stretch reflex plays an important part in the maintenance of the upright posture [p. 128].

The stretch reflex is the simplest of all spinal reflexes, in that it involves only two neurones and one synapse.

There are more elaborate spinal reflexes such as the withdrawal reflex. If a noxious stimulus is applied to one foot, that leg is withdrawn whilst the other leg is extended. Such a spinal reflex is normally suppressed by the **inhibitory reticulospinal** pathways which form part of the extrapyramidal system. The withdrawal reflex is seen after a spinal transection has interrupted these inhibitory pathways.

Extrapyramidal System. In addition to the cortico-spinal tract for voluntary movement, there are nerve fibres coming from the suppressor band area (areas 4 S and 6) which lie just in front of the motor cortex. This nerve fibre tract picks up contributions from the mid-brain, the cerebellum, and the medulla to run down to act on the anterior horn cells of the spinal cord. These fibres form part of the extrapyramidal system.

Final Common Path

The anterior horn cell is being bombarded by many neurones, some inhibitory, some excitatory, at the same time. The corticospinal tract is the principal excitatory (plus) nerve which brings about voluntary movement. The inhibitory reticulospinal tract is the principal inhibitory (minus) nerve. It inhibits activity of the anterior horn cell. It is not powerful enough to inhibit the corticospinal tract for voluntary movement, but it is powerful enough to suppress most of the spinal reflexes.

Other parts of the extrapyramidal system include the vestibulospinal tracts which originate in the vestibular nuclei.

Algebraic summation occurs at the anterior horn cell [Fig. 164]. The pluses and the minuses cancel each other out, and the discharge frequency of the anterior horn cell depends on the net result. The resultant activity is transmitted to the motor unit along the motor neurone which is thus the **final common path.**

Inhibition of Spinal Reflexes

The presence of the inhibitory reticulospinal tract makes it difficult clinically to demonstrate reflexes. The knee-jerk is one of the few spinal reflexes that can be demonstrated in most patients. Some patients do not show a knee-jerk because this spinal reflex is being suppressed by the inhibitory reticulospinal activity. In such a case, the knee-jerk can often be demonstrated whilst the subject is concentrating on some other movement such as pulling the clasped hands apart. This is termed **reinforcement.**

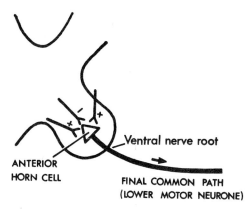

FIG. 164. The anterior horn cell in the grey matter of the spinal cord is under the influence of many neurones, some excitatory (+) and some inhibitory (−). Algebraic summation takes place with the minuses cancelling out the pluses. If the net plus activity is sufficient, the anterior horn cell discharges down the lower motor neurone to the motor unit. This nerve is termed the final common path.

Patients with brain damage or spinal cord lesions which interrupt the inhibitory pathways have exaggerated spinal reflexes. The exaggeration of the stretch reflexes gives rise to **spasticity.** A child may be a spastic as a result of a cerebral haemorrhage at birth which has destroyed these nerve pathways.

In a normal person a joint, such as the elbow joint, can be passively flexed or extended quite easily. When repeating the same manoeuvre on a spastic whose spinal reflexes have not been suppressed, as soon as one starts to flex the elbow joint, the triceps will be stretched and this muscle will promptly contract.

If sufficient force is used receptors in the tendons will inhibit this stretch reflex so that flexion suddenly occurs. This is termed the **clasp-knife reflex.** It is probably a protective mechanism to prevent rupture of the tendon.

This spasticity makes voluntary movement very difficult. The effect is usually unilateral since the brain damage is only on one side.

CEREBELLUM

The cerebellum has three main functions. The **flocculonodular lobe** of the cerebellum is associated with the vestibular apparatus and its destruction leads to disturbances of balance. A child with a medulloblastoma brain tumour which affects this lobe will be unsteady when trying to stand or walk.

The **anterior lobe** of the cerebellum is associated with proprioception. Nerve impulses from the proprioceptors reach this part of the cerebellum via the spinocerebellar tracts. The majority of these fibres do not cross and thus reach the cerebellum of the same side. This part of the cerebellum plays an important part in regulating muscle tone.

The **rest of the cerebellum** co-ordinates movement in association with the cerebral cortex of the other side. It sends information to this cortex via the thalamus, and receives information from this cortex via the pons.

A patient with a **cerebellar lesion** on one side shows a reduction in muscle tone on that side so that the hand becomes flail-like and does not swing when walking. The lack of muscle tone in the leg of the affected side may cause him to fall to that side and he may be taken for a drunken man.

The movements are poorly co-ordinated and often the incorrect force is used. In addition there is an intention tremor which becomes worse as the subject concentrates on the movement, so that even simple movements such as drinking a cup of tea become impossible, the tremor has emptied the cup before it reaches the mouth.

SETTING OF MUSCLE SPINDLES

The stretch reflex tends to hold a muscle at a given length no matter how great the force pulling on it. If this force increases, the muscle will tend to be stretched to an increased length. But any such stretching will increase the firing of the muscle spindles and reflexly increase the contraction of the muscle so that it once again matches the opposing force. If stretching force on the muscle is reduced and the muscle starts to shorten, the muscle spindle activity will be reduced and the muscle tension will decrease until it once again matches the opposing force.

A given setting of the muscle spindle thus corresponds to a given length of the muscle. The muscle spindles are adjusted to the correct setting for the maintenance of a given posture by the activity of small diameter motor nerves (gamma efferent activity) which supply intrafusal muscle fibres in the muscle spindles themselves. This activity is regulated by the extrapyramidal system. It appears to originate mainly from the cerebellum.

BASAL GANGLIA

The basal ganglia is the name given to areas of grey matter at the base of the cerebral cortex. The constituent parts are the globus pallidus, the corpus striatum, the substantia nigra and the subthalamic nucleus.

Parkinson's disease (paralysis agitans) is a disorder of the basal ganglia. It is associated with an increase in muscle tone in both the extensor and flexor muscles so that the patient, although not paralysed, has difficulty in moving any voluntary muscle. This state is termed **muscular rigidity.** The patient has a fixed mask-like expression, a shuffling gait, and a pill-rolling tremor of the hands (as if he is rolling pills with his fingers). The condition may be relieved by destroying the globus pallidus.

OVER-ALL PATTERN OF NERVOUS SYSTEM

Having considered the constituent parts of the nervous system, it is now possible to consider the principal pathways as an integrated whole [FIG. 165].

Sensory information, which enters the spinal cord (or brain) passes to the thalamus and thence to the sensory cortex for conscious sensation. It passes to the reticular formation to alert the cortex and to the cerebellum for the co-ordination of movement and the regulation of muscle tone. It passes directly to anterior horn cells in the same or adjacent segments for spinal reflexes.

The anterior horn cell, which is controlling the activity of its motor unit via the lower motor neurone,

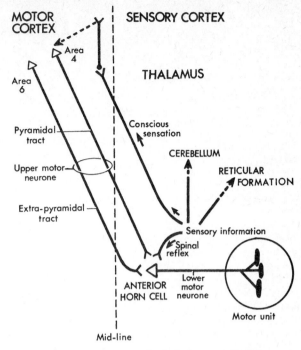

FIG. 165. Summary of principal nerve pathways.

is under the influence of the corticospinal (pyramidal) tract from the motor cortex, and the extrapyramidal tract (from area 6 of the cortex and from the brain stem) and local spinal reflexes. The pyramidal and extra-pyramidal tracts constitute the upper motor neurone. The spinal reflexes are for the most part inhibited by the inhibitory reticulospinal component of the extra-pyramidal system. Following an upper motor neurone lesion, this inhibition of the spinal reflexes is lost and the reflexes become exaggerated causing **spasticity.** If the pyramidal tract is also interrupted there will be a **spastic paralysis.** Spasticity may also be caused by excessive gamma efferent activity to the muscle spindles.

If the lower motor neurone pathway is interrupted there will be a **flaccid paralysis.**

21. THE EYE

FOCUSING

The cornea is transparent and allows light to enter the eye. It has no blood vessels and receives its nutrition from the aqueous humour.

The front surface of the cornea acts as a *fixed* lens and the optics of the eye produce an inverted image on the **retina** [FIG. 166]. The retina converts this inverted

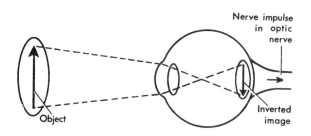

FIG. 166. The cornea and lens of the eye act as a lens system which produces an inverted image on the retina. The brain turns this inverted image upright.

picture into nerve impulses. These nerve impulses are transmitted along the optic nerves to the occipital lobes of the brain where they are interpreted as sight.

When using a camera a different setting of the focus is needed when photographing a close object to that used when photographing a distant object. In a similar way, inside the eye an *adjustable* lens is needed in order to ensure that the image on the retina is sharply in focus, no matter how far away the object viewed.

When viewing a close object such as when reading a book, this lens in the eye itself is thick and strong optically, whereas when looking at an object in the distance, the lens in the eye is thin and weak. It is this change in the thickness of the lens that enables focusing to occur.

The lens is an elastic structure and has ligaments attached around its edge. These ligaments connect the lens to the **ciliary body.** It is the pull of these ligaments that flattens the lens for distant vision [FIG. 167(B) and (C)].

The ciliary muscle which runs in an annular fashion in the ciliary body is supplied with parasympathetic nerve fibres from the third cranial nerve. For near vision nerve impulses cause contraction of the ciliary muscle fibres, and this contraction relaxes the pull of the ciliary ligaments [FIG. 167]. The lens reverts to its thicker state [FIG. 167(A)].

Should the ciliary muscles become paralysed, a condition termed **cycloplegia,** focusing no longer occurs, and it is no longer possible to read a book.

Ability of the Eye to Alter its Lens Power

A child is able to change the thickness of his lens so as to read at a distance as close as four inches, or to see an object in the far distance. With age, the ability to focus close objects is lost, and the **near point** (the closest point that can be seen clearly) recedes. By middle age the near point has receded to such a distance that additional lenses may have to be placed in front of the eyes in order to bring the image sharply into focus on the retina. Spectacles are then used for reading.

The lens may become opaque with age and this is termed a **cataract.** This leads to blindness. The opaque lens is removed at a cataract operation, but the cornea is still present to act as a lens and with an additional strong lens in front of the eye to make up for the loss of the lens in the eye, sight is once more possible.

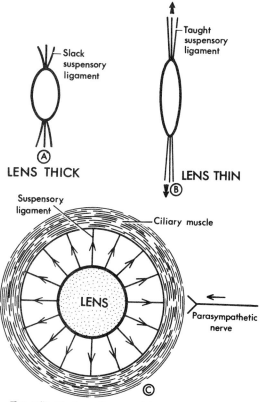

FIG. 167. The lens, if removed from the eye, would be thick (A). In the eye when looking at a distant object, the lens is thin because of the tension of the suspensory ligaments (B) and (C). When focusing to view a close object activity in the parasympathetic nerve causes contraction of the ciliary muscle. This relaxes tension on the suspensory ligaments and the lens reverts to its thick shape. It now becomes a more powerful lens and brings the object into focus on the retina.

Short Sight, Long Sight and Astigmatism

In a short-sighted person the curvature of the front surface of the cornea is incorrect, and when looking at a distant object the image falls in front of the retina. To correct this, a diminishing lens (concave) is introduced in front of the eyes to allow the object to be sharply in focus. Near objects, however, can be brought into sharp focus on the retina by a short-sighted person without the need for spectacles. Short-sightedness is termed **myopia** [FIG. 168(B)].

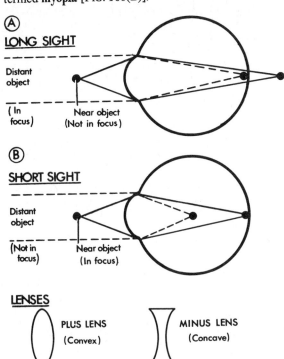

FIG. 168. A long-sighted person (A) has an eye in which, even with maximum thickness of the lens, a near object is brought to focus behind the retina. An additional 'plus' lens (in the form of spectacles) will be needed to see this object clearly.

A short-sighted person (B) can bring a near object into focus on the retina but the image of a distant object falls in front of the retina. A minus lens will be required for a distant object to be seen clearly.

Dotted lines = light pathway from distant object, solid lines = light pathway from near object.

With a long-sighted person the lens power is insufficient, and a convex lens is needed in order to bring near objects into sharp focus. This is termed **hypermetropia** [FIG. 168(A)].

If the cornea is not truly spherical, but is slightly cylindrical, horizontal lines will be focused with a different setting of the lens in the eye to that used when viewing vertical lines. As a result it will be impossible to see both clearly at the same time. This condition is termed **astigmatism.** It is corrected by introducing an equal and opposite cylindrical lens in front of the eye to cancel out the defect.

As an alternative to spectacles, contact lenses may be employed. These are applied to the front surface of the cornea, and change the curvature of this front surface to the correct shape. Contact lenses may also be employed to improve vision in people with normal vision, but who have irregularities in the front surface of the cornea as a result of trauma.

If a contact lens covers only the cornea, it is termed a *corneal lens*. If it is larger and covers the whole of the visible area, it is termed a *scleral lens*. Scleral lenses float on a thin layer of fluid (tears) and can be kept in place for 12 hours. Apertures in the lens allow the gaseous exchange of oxygen and carbon dioxide to take place.

It will be seen that it is the front surface of the cornea that is important for sight, and that the lens is only of secondary importance in that it allows the focus to be changed from a distant object to a near object.

If the eyes are opened when swimming under water, vision is blurred even though the water is clear. It is not possible to see clearly, because the water has a similar refractive index to the cornea. As a result it no longer acts as a lens. In order to act as a lens the cornea must have air in front of it. If a face-mask or goggles are used so as to trap air in front of the eye, then it is possible to see clearly during underwater swimming.

Dioptres

Optical lenses worn in front of the eyes to correct vision defects are prescribed in units known as *dioptres* (D).

A lens which will bring a distant object to a focus at a point 1 metre behind the lens is said to have a power of 1 dioptre. Another lens which will do so at $\frac{1}{2}$ metre is said to have a power of 2 dioptres. A lens which will do so at $\frac{1}{3}$ metre has a value of 3 dioptres, and so on.

The dioptre value of a lens is thus the reciprocal of its focal length in metres.

$$\text{Dioptre power} = \frac{1}{\text{focal length in metres}}$$

$$= \frac{100}{\text{focal length in cm.}}$$

The human eye brings a distant object to a focus on the retina at a distance of 1·69 cm. behind the lens system. The lens system thus has a power of $100/1·69 = 59$ D. 43D of this is provided by the curvature of the cornea, and 16D by the lens when focused at infinity (43D + 16D = 59D). The lens increases by 3D when reading a book at $\frac{1}{3}$ metre making a total of 62D.

Young children are able to increase the power of their eyes by +11D. This enables the focus point to be brought in from infinity to $\frac{1}{11}$ metre = 9 cm. Such children can read a book at this distance from their eyes.

The ability to increase the dioptre power of the eye for close reading declines with age, and by 50–60 years of age has fallen to only +1D. The nearest point of sharp focus is then 1 metre (and it would need extremely long

arms to be able to hold a book or newspaper at this
distance!). At the normal reading distance of approximately $\frac{1}{3}$ metre an increase of +3D is needed. Since
the eye can then only provide +1D of this increase, an
additional +2D is needed in the form of reading
spectacles. For close work higher powers will be needed.

A long-sighted person has a natural lens system in his
eye which is too low in dioptre power, and his need is
for plus lenses (convex) to increase the dioptre power.
Suppose, for example, that the maximum power is
59D. This is all right for distant vision, but a +3D lens
would be needed in the form of spectacles to read a book.

A short-sighted person has an eye with too high a
dioptre power. He needs negative lenses (concave) to
reduce this dioptre power. Consider a short-sighted
person with a minimum dioptre power of 62D. This is
all right for reading a book, but he needs a −3D lens to
correct his defect for distant vision (62D − 3D = 59D).

So far only spherical lenses have been considered.
For the treatment of astigmatism cylindrical lenses are
required. These lenses have a dioptre power in one
direction only. At right angles to this direction the
dioptre power is zero. The prescription for cylindrical
lenses for astigmatism will include not only the dioptre
value of the cylinder, but also the angle at which it is to
be mounted with respect to the horizontal. Correction
for astigmatism requires that the lens be mounted at the
prescribed angle; for this reason, contact lenses which
tend to rotate when worn, prove unsatisfactory for the
treatment of astigmatism. The majority of astigmatics
also need spherical correction as well.

THE PUPIL OF THE EYE

The amount of light entering the eye is controlled by
the iris diaphragm. The circular aperture is known as
the **pupil.** The iris lies between the cornea and the lens
[FIG. 169].

The size of the iris diaphragm regulates the amount of
light entering the eye (just as it does in the case of a
camera). In addition a small aperture enables both near
and distant objects to be brought into focus and seen
clearly at the same time. This is termed a *great depth of
focus*. A small aperture also reduces distortion and
produces a sharper image.

The size of the pupil is regulated by the circular and
radial smooth muscle fibres in the iris itself. These
muscle fibres are supplied by the sympathetic and parasympathetic nervous systems.

The parasympathetic fibres constrict the pupil. These
fibres run in the third cranial nerve.

The sympathetic fibres dilate the pupil. They arise
in the first thoracic segment and relay in the superior
cervical ganglion.

The smooth muscle controlling the size of the pupil
can be seen in action. This is one of the few places in
the body where the contraction of smooth muscle can
be seen. Stand in front of a mirror, cover your eyes with
your hands. The darkness will cause the pupils to
dilate. Remove your hands and follow the changes in

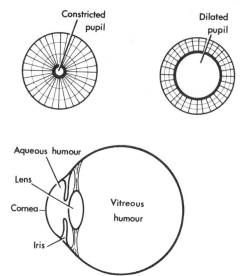

FIG. 169. The chamber in front of the lens is filled with
aqueous humour whilst that behind the lens is filled
with vitreous humour. The amount of light entering
the eye is adjusted by the iris diaphragm. This contains
circular and radial smooth muscle fibres. Increased
sympathetic activity gives a dilated pupil. Increased
parasympathetic activity gives a constricted pupil
[see FIG. 135, p. 118].

the size of the pupil. It will constrict fairly slowly and
take up a new smaller size.

The method employed clinically to test the **pupillary
reaction** is to shine a light in the eye using a torch. The
resultant constriction of the pupil is termed the **light
reflex.** The two pupils work together so that shining a
light into one eye causes both this pupil and the pupil
of the other eye to constrict (**consensual reflex**).

Constriction of the pupil occurs in bright light, when
looking at a near object, and when there is excessive
parasympathetic activity to the eye [p. 117]. Dilatation
of the pupil occurs in darkness, and when there is
excessive sympathetic activity, such as in emotional
excitement.

The size of the pupil depends on the balance between
the sympathetic and parasympathetic activity [FIG. 135,
p. 118]. Thus constriction of the pupil may be due to
excessive parasympathetic activity or due to insufficient
sympathetic activity. Morphine, for example, produces
a pin-point pupil because it stimulates parasympathetic
activity.

A dilated pupil may be due to excessive sympathetic
activity or insufficient parasympathetic activity. Eye
drops for dilating the pupil contain **homatropine** (which
is a derivative of *atropine*). It blocks the parasympathetic activity. The unopposed sympathetic activity
dilates the pupils.

CONVERGENCE

When looking at a distant object, the two eyes are
looking in directions that are parallel to one another.
As one looks at a closer object, the eyes turn inwards so

that both eyes can look at the same object. This moving in of the visual axes is termed **convergence**. It is brought about by the extrinsic muscles which control the movements of the eye, and particularly by the internal (medial) rectus muscles which medially rotate the eyes.

Convergence Reflex

The eyes at rest are focused at infinity. When we look at a close object three things happen:

1. The eyes converge.
2. The pupils constrict.
3. The increase in thickness of the lens enables the object to be sharply focused.

Should the two eyes not point in the same direction the condition is termed a **squint.**

Small discrepancies between the visual axis may be corrected using wedge-shaped lenses in front of the eyes.

AQUEOUS AND VITREOUS HUMOURS

The chambers in the eye in front of the lens are filled with a watery fluid known as **aqueous humour.** The chamber behind the lens is filled with a jelly-like fluid known as **vitreous humour.**

Aqueous humour is secreted by the **ciliary process** of the ciliary body which is situated in the **posterior chamber** between the iris and the lens. The aqueous humour passes through the opening in the iris and enters the **anterior chamber** between the iris and the front of the

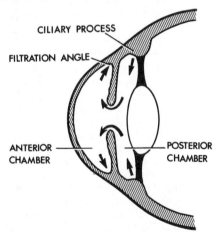

FIG. 170. Formation and reabsorption of aqueous humour. Aqueous humour is secreted by the ciliary process in the posterior chamber. It passes through the aperture in the iris to be reabsorbed in the filtration angle in the anterior chamber. Note that both the anterior and posterior chambers are in front of the lens.

eye [FIG. 170]. It is absorbed in the **filtration angle** into the canal of Schlemm which leads to the veins.

The pressure of the aqueous humour is 25 mm. Hg. If the formation of aqueous humour exceeds the reabsorption, then the pressure inside the eye will increase. This condition of raised intra-ocular tension is termed

glaucoma. It requires immediate treatment to prevent the sight being permanently impaired.

Care must be taken when giving drugs which dilate the pupil to a person who has suffered from glaucoma, since the dilatation may obstruct the filtration angle where reabsorption occurs. It may thus precipitate a further attack of glaucoma.

RETINA

The light sensitive elements of the retina consist of rods and cones [FIG. 171]. The cones are used during the day for colour vision. The rods are used for night vision and give only the sensation of shades of grey. We are thus colour blind when using the rods.

The central part of the retina is tightly packed with cones. The image is brought to a focus at this part of the retina when, for example, we are reading a book.

When looking at stars on a dark night, it is often possible to see a faint star more clearly out of the corner of the eye than when looking straight at it. This is due to the presence of rods in the peripheral vision.

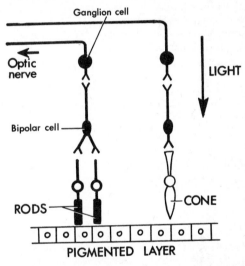

FIG. 171. Structure of the retina. The cones are used for daylight colour vision. The rods are used for monochrome night vision. It will be noted that the light has to pass through the ganglion cells and bipolar cells to reach the rods and cones.

Ophthalmoscope

When the eye is completely relaxed, it is focused at infinity and parallel rays of light are brought to a sharp focus on the retina. If both the observer and the subject have their eyes focused at infinity, then it is possible for the observer looking directly at the subject to see the subject's retina sharply in focus. The inside of the subject's eye is, however, dark and must be illuminated before any detail can be seen.

An ophthalmoscope is basically an apparatus for illuminating the retina using a battery and small bulb

[FIG. 172]. For normal vision the ophthalmoscope consists simply of a small hole to look through and a source of illumination. Lenses are provided to correct focusing defects when either the observer or the subject normally wears spectacles.

The macula is seen as a small oval area devoid of blood vessels. It is a deeper red than the rest of the retina and has a whitish centre (the fovea). This is the point of sharpest vision.

The blood vessels and the optic nerve enter at the optic disc. The central optic artery usually has a central stem which divides into two branches at 180 degrees. The central vein usually has two trunks and is lateral to the artery. The arteries in the retina are narrower and a brighter red than the veins, and they have a well marked light streak along the middle. This is a reflection of the ophthalmoscope light. The veins pulsate, the arteries normally do not.

Normally the light streak is lost as the vessels pass over the edge of the optic disc. In **papilloedema** the disc swells and the vessels bend over its edge. The light in this region is not reflected back, so the bent portion of the vessel appears dark. In **hypertension** the veins are nipped by the arteries as they cross. The arteries are narrowed and look like silver wire. There may be exudates that look like cotton wool.

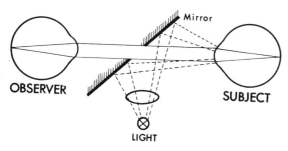

FIG. 172. Principle of the ophthalmoscope. When the eye is focused at infinity, parallel rays of light are brought to a focus on the retina. If both the observer and the subject focus at infinity they will theoretically be able to see each other's retinae, the rays of light being parallel between the two eyes. In practice the inside of the observer's eye is too dark for the retina to be seen whilst the subject's retina is illuminated. In the form of ophthalmoscope illustrated, the light is reflected into the eye by a mirror. The observer looks directly through a hole in the centre of the mirror. The ophthalmoscope has lenses to correct for visual defects of either the observer or subject.

Colour Vision

When white light is separated out into its component colours a spectrum or rainbow is obtained. Although the eye is sensitive to all these colours, the sensation of colour can be obtained using only three colours, red, green and blue. This principle is used in a colour television set and in colour photography, where only three colours are used to produce the entire spectral range.

The Young-Helmholtz theory of colour vision is based on the presence of three different types of cones which are sensitive to these three primary colours. Such a theory explains the various types of colour blindness. A person who cannot see red as a colour has a condition termed **protanopia**. If green cannot be seen, the condition is termed **deuteranopia**. Occasionally blue cannot be seen and this is termed **tritanopia**; it is a very rare condition.

Colour blindness may be detected using Ishihara colour blindness test charts which have been designed so that a colour-blind person sees a different figure or pattern to a normal person.

Purkinje Shift. At dusk there is a change over from using cones to using the more sensitive rods. This results in the blue colour appearing unusually bright compared with the red colour. This can be seen in the case of a plant such as the geranium where the red flowers become very dark as the light intensity decreases. It is due to the fact that the rods are most sensitive in the blue-green region whereas the cones are most sensitive in the yellow-green region of the spectrum.

The change from **photopic** vision (cones) to **scotopic** vision using entirely rods takes about 30 minutes in complete darkness. This change from light-adapted vision (photopic vision) to dark-adapted vision (scotopic vision) is called dark adaptation. It is impaired when there is a deficiency of vitamin A since a derivative of vitamin A (vitamin A—aldehyde) forms the pigment visual purple which is essential for night vision.

Binocular and Monocular Vision

The use of two eyes for vision has many advantages over the use of only one eye. Two eyes give a larger visual field, and a defect in part of the field covered by one eye will not be noticed, because this will be covered by the other eye. This may be illustrated in connexion with the blind spot. We are blind in each eye over the area of the retina where the optic nerve enters (optic disc), but we are unaware of this defect since the other eye will see objects in this area.

Binocular vision gives a slightly different retinal picture in each eye, and this gives perspective and a sense of size and shape to objects. It is termed stereoscopic vision. The effort of convergence and the stereoscopic effect enables the distance of objects to be judged. With only one eye the judgement of distance is greatly impaired.

Visual Field Defects

Although the nerve impulses from the fovea of each eye pass to both the visual areas of the brain, the outer fields pass to only one side of the brain. That is, they are unilaterally represented. As a result, lesions of the optic pathways may result in the loss of a peripheral part of the visual field. This may be detected using a perimeter.

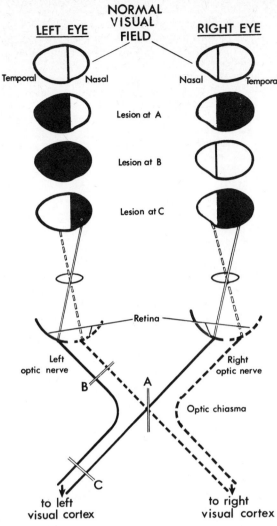

FIG. 173. The visual fields are determined by a perimeter. The light crosses over as it enters the eye so
that the light from the lateral field falls on the medial
part of the retina, whilst light from the nasal field falls
on the lateral part of the retina. Lesions at A, B and C
produce different visual field defects.

Normally when looking straight ahead it is possible
to see objects that are more than 90 degrees to one side,
that is, slightly behind you on the two lateral sides. On
the nasal side the field of each eye is limited by the nose.

These fields of view are greatly curtailed if the nervous
pathways are interrupted. The type of visual loss can
be used to determine where the lesion has occurred.

A lesion at **A** [Fig. 173], which might be due to a
pituitary tumour pressing on this part of the optic nerve,
will interrupt nerve pathways from the medial sides of
the two retinae. These areas of the retina are used for
peripheral vision. The patient will lose the lateral parts
of the visual field giving an effect similar to wearing
blinkers.

A lesion at **B** will cause blindness in one eye, leaving
normal vision in the other.

A lesion at **C** will result in the loss of vision by both
eyes on one side of the body. If the lesion is on the left
side as in FIGURE 173(C) the patient will be able to see
straight ahead with a normal left visual field, but the
right side of the visual field will be blacked out.

Visual Acuity

If a picture in a newspaper is examined under a
magnifying glass, it is seen to consist of a series of dots.
With the naked eye these dots cannot be seen. This is
because the visual acuity of the eye is not great enough
to discriminate between the individual dots.

The normal eye can see two points as separate dots
if they subtend an angle of at least 1 minute (one sixtieth
of a degree) at the eye.

The **acuity** is defined as 1 divided by the angle
expressed in minutes. Thus if the minimum angle is 1
minute, the acuity is one $\left(\frac{1}{1} = 1\right)$, whereas if the minimum angle is 2 minutes the acuity is $\frac{1}{2} = 0\cdot5$.

Test types are employed to determine visual acuity.
They consist of black letters on a white background of
such a size and shape that the detail of each letter will
subtend an angle of 1 minute if the chart is viewed from
the distance specified for each line of type.

For distant vision these types were originally designed for a viewing distance ranging from 5 metres to
60 metres. The standard distance for viewing this type
is now always 6 metres. If the 6 metre line can be read,
the vision is said to be $\frac{6}{6}$. If the smaller 5 metre line of
type can be read from the standard distance of 6 metres
the acuity is said to be $\frac{6}{5}$. If only the larger 18 metre
type can be read at this distance, the acuity is only $\frac{6}{18}$.

22. THE EAR

Sound consists of vibrations in the air. These vibrations vary in frequency from 20 Hertz (cycles per second) for the deep notes up to 10,000–20,000 Hz (cycles per second) for the very high notes.

The **external ear** collects the sound waves and transmits them to the ear-drum (tympanic membrane) [FIG. 174].

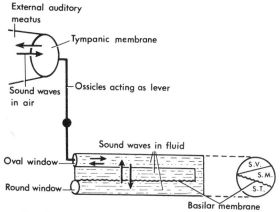

FIG. 174. Diagram showing mode of action of the ear. Sound waves arrive at the tympanic membrane (ear-drum) as vibrations in the air. The ossicles act as a lever and convert the large but weak vibrations of the ear-drum into small but powerful vibrations of the oval window. When the oval window moves in, the pressure is transmitted through the scala vestibuli (S.V.) to the scala media (S.M.) and then to a basilar membrane and the scala tympani (S.T.). The round window bulges outwards. The vibrations of the basilar membrane produce nerve impulses in the auditory nerve.

Three small ossicles in the **middle ear** (on the inner side of the ear-drum) act as a lever and transmit the sounds to the oval window of the **inner ear.** The inner ear is filled with a watery fluid, and this arrangement allows the large but weak vibrations of the air to be changed into smaller but more powerful vibrations in the fluid. This process is termed *impedance-matching*.

The middle ear is air-filled and communicates with the nasopharynx via the **auditory tube** (Eustachian tube). This tube is normally closed, but it opens during swallowing. When open the pressure on the inside of the ear-drum becomes the same as the pressure outside. This is important because sound will only make the ear-drum vibrate correctly if pressures on the two sides are equal. If the pressures are unequal, deafness will result.

A sudden change in the atmospheric pressure will cause deafness until the pressures have equalized again by swallowing. During the take-off of a plane, there is often a drop in external pressure as the altitude increases and unless the auditory tube is opened frequently discomfort and deafness will result. Children who are too young to understand this, may suffer intense ear-ache due to this cause. Giving the child a sweet to suck is one way of ensuring frequent swallowing.

If the auditory tube becomes blocked, for example, with the mucus associated with a cold in the head, the air in the middle ear is absorbed into the blood and the ear-drum will be drawn inwards leading to deafness. This deafness disappears when the auditory tube block is removed.

INNER EAR

The vibrations in the fluid of the inner ear are converted into nerve impulses in the **cochlea.**

The cochlea consists of a series of tubes wound in a $2\frac{3}{4}$-turn spiral round a central pillar. The whole structure resembles a snail's shell hence the name (L. *cochlea* = snail).

The oval window leads to the first of these tubes known as the **scala vestibuli** (*scala* (L.) = spiral staircase). This is filled with a watery fluid termed **perilymph** which is very similar in composition to cerebrospinal fluid.

The scala vestibuli is separated by means of a membrane from the **scala media.** The scala media is filled with **endolymph** which is a fluid similar to that found in cells, and has a high potassium concentration.

The scala media is separated from the third tube, the **scala tympani,** by the **basilar membrane.**

The sound vibrations in the fluid of the scala vestibuli are transmitted through to the fluid in the scala media, the basilar membrane and to the fluid in the scala tympani [FIG. 175]. When the oval window moves in, the round window moves out and vice versa.

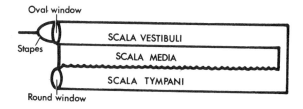

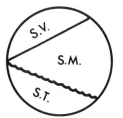

FIG. 175. The scala vestibuli, scala media and scala tympani are wound round a central pillar forming $2\frac{3}{4}$ turns of the cochlea.

It is the vibration of the basilar membrane that produces nerve impulses in the auditory nerve. The basilar membrane is short at its base near the oval window, and increases in length to reach a maximum size at the apex of the spiral. The nerve endings are found at the base of the **hair cells** of the **organ of Corti**. This lies on the basilar membrane [FIG. 176]. The hairs

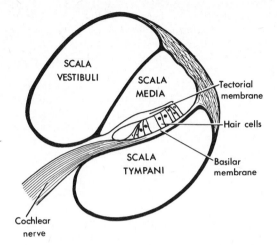

FIG. 176. Basilar membrane. The cochlear nerve has sensory receptors in the hair cells on the basilar membrane. The hairs are embedded in the tectorial membrane. These hair cells generate the auditory nerve impulses.

themselves are embedded in the **tectorial membrane**. Low frequency sounds cause the whole of the basilar membrane to vibrate. High frequency sounds cause only the base of the basilar membrane close to the oval window to vibrate.

At low frequencies a nerve impulse is produced for every vibration of the basilar membrane. The sense of pitch is thus conveyed by the frequency of the nerve impulse.

At higher frequencies only part of the basilar membrane is stimulated by sound, and the sense of the pitch is conveyed by the pattern of vibration of the basilar membrane.

The nerve impulses pass via the cochlear division of the eighth cranial nerve to the cochlear nuclei. Here the first neurone stops and the second neurone starts. The second neurone crosses to the other side by the lateral lemniscus and runs to the inferior colliculus, and passes to the medial geniculate body. Here the second neurone stops and the third neurone runs from the medial geniculate body via the posterior limb of the internal capsule to the auditory cortex (superior temporal gyrus).

DEAFNESS

Deafness due to disease of the external auditory meatus or the middle ear is termed **conduction deafness**. The common causes are: wax blocking the external auditory meatus, damage to the ear-drum and failure of the ossicles to transmit the sound from the ear-drum to the fluid in the scala vestibuli. In otosclerosis, the stapes may have become fixed. This condition is treated by replacing the stapes by a plastic prosthesis, and the oval window by a graft (from a vein). This operation is performed using a binocular microscope.

Since sound can reach the inner ear via the bones of the skull, conduction deafness is never complete. In a normal person, the sounds heard by bone conduction become louder when the **masking** effect of room noise is stopped by using an ear-plug. This fact forms the basis of the Rinne's and Weber's tests.

Deafness due to disease of the auditory nerve is termed **nerve deafness**. Deafness in this case may be absolute.

Rinne's and Weber's Tests

When carrying out **Rinne's test**, a tuning fork is placed on the mastoid process until it can no longer be heard by bone conduction. It is then placed close to the ear and in a normal subject, it will be heard by air conduction for a further length of time (Rinne positive). In conduction deafness this is not so (Rinne negative). In the normal subject, the tuning fork ceases to be heard by bone conduction, although it is still vibrating, because of the masking effect of the room noise which is arriving by air conduction. In the case of conduction deafness, this room noise will not be heard.

With **Weber's test** the tuning-fork is placed on the top of the head. Normally the sound is heard equally in the two ears, but if there is unilateral conduction deafness, it will be heard better in that ear. This is because the masking effect of the room noise will be absent on that side. As a memory aid it may be noted that **W** for Weber is a symmetrical letter, and that in Weber's test the tuning fork is placed symmetrically with respect to the head.

Decibels

The loudness of sound is measured in decibels. Hearing is most acute at about 1,000 cycles per second (= 1,000 hertz), at which frequency a sound of intensity 0·002 dynes per square centimetre (2×10^{-5} Newtons per square metre) can be heard under ideal conditions. This is taken as the start of the **decibel scale** and is termed 0 db. The decibel scale is a logarithmic scale where every increase of 10 decibels means an increase of 10 times in sound power. Thus 20 decibels is an increase of 100 times. 30 decibels is an increase of 1,000 times and so on.

Decibels	
0	Threshold of hearing
40	Talking in whisper
80	Traffic noises
120	Pneumatic drill
140	Jet plane at take-off

Sounds over 120 decibels produce pain in addition to sound.

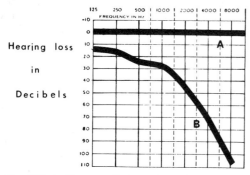

FIG. 177. Audiogram. Each ear is tested to find the minimum sound level at which the different frequencies can be heard. Left ear (A) has normal hearing. Right ear (B) has high frequency deafness.

Audiometer

The hearing is tested at different frequencies using an audiometer. This consists of an electronic oscillator which is connected to a pair of specially calibrated headphones. With this apparatus a note of any pitch and loudness can be applied to either ear of the subject who indicates whether or not the sound can be heard. By finding the minimum intensity of sound that can be heard at each frequency, the subject's hearing can be compared with the normal and the hearing loss determined. Ideally any hearing-aid provided should match this hearing loss.

As has been seen, the ear is most sensitive to frequencies around 1000 Hz. Its sensitivity decline as the frequencies fall towards 20 Hz and rise towards 20,000 Hz. This fact is taken into account when designing the audiometer. Low and high frequencies are boosted so that all frequencies appear to have the same intensity to the normal ear, and the apparatus is adjusted so that all frequencies can just be heard when the volume control is set at 0 db.

If a patient is deaf, a higher setting of the volume control will be needed for a particular frequency to be heard. This volume control is calibrated in decibels above the normal.

By testing the hearing of each ear at different frequencies and plotting a graph, an audiogram is obtained. FIGURE 177 shows the audiogram of a patient with normal hearing in the left ear (A) but with high frequency deafness in the right ear (B). With this ear a higher volume is needed for sounds to be heard, and at the really high frequencies, sounds are not heard at all.

VESTIBULAR APPARATUS

The inner ear also contains the three semicircular canals and the otolith organs, the saccule and the utricle. The semicircular canals respond to acceleration whilst the saccule and utricle respond to the position of the head in space [see p. 129].

23. SENSATIONS OF TASTE AND SMELL

SMELL

The sense of smell originates in the **olfactory epithelium** which is found in a very small area of mucosa high in the nasal cavity in man. This area lies above the main air stream, and the normal air breathed in and out by respiration by-passes this region. The sense of smell is aroused by eddy currents which carry an odoriferous substance to the receptors. A sniff directs the air stream directly to the receptors.

The nerve supply of this olfactory mucosa is the olfactory nerve which is the first cranial nerve (I).

The sense of smell shows rapid adaptation, and some odours will mask others.

TASTE

Four types of taste receptors known as **taste buds** are found on the upper surface of the tongue. These four sets of receptors give the sensations of salt, sweet, sour and bitter. A salt taste is given by a substance such as sodium chloride; a sweet taste is given by a substance such as sugar; a sour taste by an acid such as hydrochloric acid; and a bitter taste by a substance such as quinine.

The receptors lying at the front of the tongue (anterior two-thirds) are supplied by the lingual nerve which is a branch of the fifth cranial nerve (V). Although this fifth cranial nerve supplies this part of the tongue with general sensation, the taste fibres leave it and pass via the chorda tympani nerve to join the facial nerve (seventh cranial nerve). The receptors lying at the back of the tongue (posterior one-third) are supplied by the glossopharyngeal nerve which is the ninth cranial nerve (IX).

Many tastes are in fact smells. Thus if the secretions produced by a cold in the head prevent stimulation of the olfactory mucosa in the nose, the taste of food is altered.

The pleasant taste and smell of food stimulates the production of saliva, gastric juice and pancreatic juice.

Taste has a protective function. Many poisons alter the taste of food and thus give away their presence.

In a complex way, the desire for certain foods is associated with the senses of taste and smell. In pregnancy there is a characteristic craving for unusual foods [p. 111]. This may have been related in our evolutionary past with a need for such foods.

When salt depleted, or when on a low sodium diet, there is a craving for foods with a salty taste. If sodium salts are not allowed, this desire can be satisfied by taking potassium chloride or ammonium chloride which have a similar but not identical salty taste.

In a similar way the desire for sweet substances, such as glucose and sucrose, can be satisfied by saccharin and the cyclamates, which are unrelated to the sugar chemically, and have no nutritional value. They are used in place of sugar when taking a low calorie diet.

INDEX